T0134856

Cognitive Systems Monographs

Volume 33

About this Series

The Cognitive Systems Monographs (COSMOS) publish new developments and advances in the fields of cognitive systems research, rapidly and informally but with a high quality. The intent is to bridge cognitive brain science and biology with engineering disciplines. It covers all the technical contents, applications, and multidisciplinary aspects of cognitive systems, such as Bionics, System Analysis, System Modelling, System Design, Human Motion, Understanding, Human Activity Understanding, Man-Machine Interaction, Smart and Cognitive Environments, Human and Computer Vision, Neuroinformatics, Humanoids, Biologically motivated systems and artefacts Autonomous Systems, Linguistics, Sports Engineering, Computational Intelligence, Biosignal Processing, or Cognitive Materials as well as the methodologies behind them. Within the scope of the series are monographs, lecture notes, selected contributions from specialized conferences and workshops.

More information about this series at http://www.springer.com/series/8354

M. Guadalupe Sánchez-Escribano

Engineering Computational Emotion—A Reference Model for Emotion in Artificial Systems

M. Guadalupe Sánchez-Escribano
Escuela Técnica Superior de Ingenieros
Industriales
Universidad Politécnica de Madrid
Madrid
Spain

Advisors:
Ricardo Sanz
Ramón Galán

ISSN 1867-4925 ISSN 1867-4933 (electronic)
Cognitive Systems Monographs
ISBN 978-3-319-86623-9 ISBN 978-3-319-59430-9 (eBook)
DOI 10.1007/978-3-319-59430-9

Printed on acid-free paper

This Springer imprint is published by Springer Nature
The registered company is Springer International Publishing AG
The registered company address is: Gewerbestrasse 11, 6330 Cham, Switzerland

Under the consent of my beloved David who has loved me and has supported all my hours of study. And the consent of my parents, my sister and my family, that has given me all their love trying to understand me…
…let me to dedicate this work to my beloved grandparents and grandmothers, who might be taking care of me from the sky.
I miss you.
To: Amalia, Agustin, Nati and Eloy

Acknowledgements

I would like to thank Ricardo Sanz for the opportunity he gave me to discover this amazing field. Without his willingness to share knowledge, resources, histories, and experience, this research would not have ever been possible. He has been an excellent friend, mentor, and teacher.

In the same vein, I would like to extend great thanks to Ramón Galán who always offered me support and comprehension in my silence. I started within the research field thanks to him, and I have to thank him everything I learnt from his expertise. My first steps in code and systems were built from the basis of working with him. He has been the perfect second advisor and mentor.

I hope to continue my research with them.

Throughout this time, one person has always been there during those difficult and trying times. She has pushed toward improving my work and offering support and help to me. I would like to express gratitude toward Paloma de la Puente, the best engineer, colleague, and friend you might find.

This text shows a piece of research that can be read under the correct understanding because of the input, influence, text improvement, and expert knowledge of Eva R. Porras. I will owe a great deal of gratitude towards you.

Most importantly of all, I show extensive gratitude to all of the people who contributed to this moment in my life. Great thanks to Jose Luis García Vidueira and Gonzalo Sánchez Casado.

Great thanks to all

Contents

List of Figures

Abstract

Emotion is generally argued to be an influence on the behavior of life systems, largely concerning flexibility and adaptivity. The way in which life systems act in response to particular situations of the environment has revealed the decisive and crucial importance of this feature in the success of behaviors. And this becomes a source of inspiration for artificial systems design.

During the last decades, artificial systems have undergone such an evolution that each day more are integrated in our daily life. The subsequent effects are related to an increased demand of systems that ensure resilience, robustness, availability, security, or safety among others. These attributes raise fundamental challenges in control and complex systems design.

This thesis has been developed under the framework of the Autonomous System project, a.k.a the *ASys Project*. Short-term objectives of immediate application focus on the design of improved systems, seeking more intelligence in control strategies. Besides this, long-term objectives underlying *ASys Project* concentrate on high-order capabilities such as cognition, awareness, and autonomy.

This work is placed within the general fields of*enginery* and*emotion science* and provides a theoretical foundation for engineering and designing computational emotion for artificial systems. The starting research question that has grounded this thesis aims at addressing the problem of emotion-based autonomy. And how to feedback systems with valuable meaning conforms the general objective. Both the starting question and the general objective have underlaid the study of emotion, the influence on systems behavior, the key foundations that justify this feature in life systems, how emotion is integrated within the normal operation, and how this entire problem of emotion can be explained in artificial systems. By assuming essential differences concerning structure, purpose, and operation between life and artificial systems, the essential motivation has been the exploration of what emotion solves in nature to afterward analyze analogies for man-made systems.

The approach provides a reference model in which a collection of entities, relationships, models, functions, and informational artifacts are all interacting to provide the system with non-explicit knowledge under the form of emotion-like relevances. This solution aims to provide a reference model as a framework to design solutions for emotional operation, but related to the real needs of artificial systems.

The proposal consists of a multi-purpose architecture that implements two broad modules in order to attend: (a) the range of processes related to the environment affectation, and (b) the range or processes related to the emotion perception-like and the higher levels of reasoning. This has required an intense and critical analysis beyond the state of the art around the most relevant theories of emotion and technical systems, in order to obtain the required support for those foundations that sustain each model.

The problem has been interpreted and is described on the basis of *AGSys*, an agent assumed with the minimum rationality as to provide the capability to perform emotional assessment. *AGSys* is a conceptualization of a *model-based cognitive agent* that embodies an inner agent *ESys*, the responsibility for performing the emotional operation inside of *ASys*.

The solution consists of multiple computational modules working federated and aimed at conforming a mutual feedback loop between *AGSys* and *ESys*. Throughout this solution, the environment and the effects that might have an influence on the system are described as different problems. While *AGSys* operates as a common system within the external environment, *ESys* is designed to operate within an*inner environment* built on the basis of those relevances that might occur inside of *AGSys*. This allows for a high-quality separated reasoning concerning mission goals defined in *AGSys* and emotional goals defined in *ESys*. This way, a possible path for high-level reasoning under the influence of goals congruence is provided.

The high-level reasoning model uses knowledge about emotional goals stability, hence allowing for new directions in which mission goals might be assessed according to the situational state of this stability. This high-level reasoning is grounded by the work of *MEP*, a model of emotion perception that is conceived as an analogy of a well-known theory in emotion science. The work of this model is described under the operation of a recursive-like process denoted *R-Loop*, together with a system of emotional goals that are assumed to be individual agents. This way, *AGSys* integrates knowledge about the relation between a perceived object and the effect which this perception induces on the situational state of the emotional goals. This knowledge might afford a new system of information that might provide the sustainability for a high-level reasoning. The extent to which this reasoning might be approached is just delineated and assumed as future work.

This thesis has been studied from a wide variety of fields of knowledge. These knowledge categories can be classified into two main area groups: (a) the fields of psychology, cognitive science, neurology, and biological sciences in order to obtain understanding concerning the problem of the emotional phenomena, and (b) a large amount of computer science branches such as autonomic computing (AC), self-adaptive software, self-X systems, model-integrated computing (MIC), or the paradigm of Model Driven Engineering among others, in order to obtain knowledge about tools for designing each part of the solution.

The final approach has been mainly performed on the basis of the entire acquired knowledge and described within the fields of artificial intelligence, Model-Based Systems (MBS), and additional mathematical formalizations to provide punctual understanding in those cases that have required it. This approach describes a reference model to feedback systems with valuable meaning, allowing for reasoning with regard to (a) the relationship between the environment and the relevance of the effects on the system, and (b) dynamical evaluations concerning the inner situational state of the system as a result of those effects.

Chapter 1
Introduction

This thesis presents a proposal of a reference model for emotion in artificial systems. The main argument is to provide a computational model of emotion appropriate for the particular purpose of an artificial system. To this end, the model has been conceived under the work of two agents in order to separate the emotional part and the rational part of the system. Thus, two agents have been proposed: the rational *AGSys*, and the emotional *ESys*. They are denoted in this way for reasons of comprehensibility. However, the emotion emerges in this proposal from the federated work between the two subagents.

AGSys is the rational agent that interacts with the external environment, under the common work of engineered systems. *ESys* is the emotional agent which interacts with the inner environment of *AGSys*, and whose rationale is aimed to reason about the state of this new environment. They are working under a mutual feedback loop which is responsible for a properly interaction.

The emotional agent *ESys* is formed by three essential elements: (a) the agent-based system of emotional goals, (b) the emotional space denoted as *Universe of Emotion*, and (c) the assessment of the emotional space by using *Emotional Attribute* vectors. The rational agent *AGSys* is based on three main parts: (a) the emotion perception by using the *Model of Emotion Perception (MEP)*, (b) the logic of emotion affectation, and (c) the conceptual work of representation.

The operation of *MEP* is conceived by using the conceptualization of a recursive function *R-Loop* and the logic of emotion affectation is devised under the work of an adaptive engine. Regarding the conceptual work of representation, it is argued the requirement of an emotion-based ontology.

An exhaustive study of the state of the art has been made in order to justify design decisions with arguments, and support these arguments with theories of relevance from emotion science. To this end, there have been proposed 61 thesis rationales which conform the set of reasons given with the aim of determining each part of the solution.

M.G. Sánchez-Escribano, *Engineering Computational Emotion—A Reference Model for Emotion in Artificial Systems*, Cognitive Systems Monographs 33,
DOI 10.1007/978-3-319-59430-9_1

1.1 Background

During the last decades, artificial systems have undergone such an evolution that we have integrated them as part of our daily life. Engineered systems increase in capabilities and abilities. The more these features grow, the more the complexity grows either be in design, development and deployment. This growth in complexity is directly related to extensive changes in the structure and dynamics of the system, since the interdependence among parts becomes more intricate as well. Our new complex systems can identify global goals and successfully solve complex problems. However, one of the main constraints in our systems is how they deal with situations that have not been minded at time design and concretely, how systems would be able to efficiently behave in complex scenarios avoiding risks. Nonetheless, there are demanded control strategies that enable systems to withstand or recover quickly from difficult conditions under sudden circumstances.

Emotion is generally argued to be an influence on the behavior of life systems, largely concerning flexibility and adaptivity. The way in which life systems act in response to particular situations of the environment, has revealed the decisive and crucial importance of this feature in the success of behaviors. And this becomes a source of inspiration for artificial systems design.

Emotion can be understood from a number of perspectives as developed in the state of the art from the various fields. Hence, to focus its contribution, this thesis had to determine purpose in using the word *emotion*. The understanding of the entire range of perspectives has been an essential requirement for the objectives of this work.

Any research question requires to study the particular condition of methods and updated research sources regarding the topic at the specific time of the investigation. The state of the art has been an essential starting point in order to solve the questions and support the later solution. This exhaustive study has been the cornerstone of this work. Not only were we needing knowledge about the state of the investigation in emotion science, but we also had to understand the essentials of the real phenomenon from each of the multiple alternative points of view in this topic.

One key objective of this work has been the analysis of biological emotion aimed to understand those principles under which biological systems optimize the relationship with their environment. We attempted to answer questions regarding the role of emotion in the adapted responses of life systems, and address a formal solution under which technical systems might implement similar capabilities.

We have maintained the vision of an artificial system as the artificial system it is, and we have allowed the artificial emotion work for itself and its own interests. This research has thus gone far beyond the issue of finding how to convey emotion to artificial agents. We have studied the state of the art in order to understand how emotions emerge, why they emerge and where are the sources from which these emotions emerge. To this end, this dissertation analyzes some of the key aspects essential to this work, bearing in mind that we do not provide an exhaustive list of all crucial issues, nor do we intend to fully analyze those included herein.

We just provide an starting point over which this field of research might follow, at the same time that we stand the framework under which we establish our research.

It is expected that artificial emotion might improve the control capabilities of systems by managing its complexity. This control aims to address tasks for self-detecting uncertain or dangerous variances inside of the system, and to avoid the progress towards non-desirable states (either if these variances have their origin inside or outside the system). That is, systems with the meta objective of well-being that analyze the patterns of objectives within the system, the state of those patterns and their immediate consequences. To this end, the system should be able to transfer phases of its control from the outside domain to its inner domain when it might be required in the form of emerging needs that can be realized by the system.

1.2 Rationale and Hypotheses

Complex technical systems inevitably force the emergence of complex engineering problems. Most of complex systems are distributed *Systems-of-Systems (SoS)* formed by different parts that cannot be functionally isolated but perform federated work and share resources (see Magee and deWeck [157]). A major challenge of systems engineering is the building of systems so they can address missions in a dependable way, regardless the operational environment or unexpected events not predicted at design time. We can regard those complex systems as Open Systems under the conception advanced by Von Bertalanffy et al. [265], i.e. systems that maintain themselves in exchange of materials with environment, and in continuous building up and breaking down of their components [265]. This produces the emergence of inner environments inside of systems, i.e. multiple states and service patterns that become a complex set of relevant circumstances that also influence the system. This concept of *environment* is a key aspect of the vision proposed in this thesis. Furthermore, the external environment in which systems perform their missions is also harsh and full of unexpected events. Global uncertainty in unstructured environments, unpredictable internal and external events, etc. cannot be fully considered in the design phases of a system construction. At design time, the information about the operation of the system is partial and, consequently, the model for controlling behavior, i.e. the acceptable changes, is incomplete. This implies that the way in which systems deal with these unplanned events strongly influences its final resilience and robustness.

This scenario reveals the need of designing novel methods to meet the requirements of systems of increasing complexity. Autonomous self-organization is a key strategy for this end. Ashby [7] explored several Principles of the Self-Organizing System arguing that "systems in general go to equilibrium" by means of the *Spontaneous Generation of Reorganization* among their multiple parts [7].

Taking a look to biology, somehow it can be argued that one essential feature of autonomous systems is their capability to maintain their *systemic equilibrium*. Their autonomy is featured by their ability to operate towards their systemic equilibrium, and activate subsequent patterns of required responses related to it. Autonomous systems are not infallible systems, but systems which operate towards their systemic stability—whether they finally achieve equilibrium or not.

Every transition from any state to one of the equilibria, requires the selection of proper states that determine the decisive stability (see Ashby [6] and [8]). The challenge of building dependable complex systems should be addressed by designing systems with the faculty of selecting those proper states, and it directly implies the design of systems with deep plasticity and self-awareness. The exploration of self-adaptive and autonomous capabilities become thus a key matter for complex critical systems design.

Generally speaking, emotion can be seen as the engine that generates meta directives towards the adaptiveness of the system into its operational environment, and this adaptiveness can be comprehended as the means to autonomously achieve systemic equilibrium. Additionally to this, emotion is generally argued as an essential piece of the intelligence (see Salovey and Mayer [226]). However, this should be correctly comprehended concerning artificial systems. An artificial system that express emotion, does not have to be an intelligent system—since it can be just a reactive system. As well, an intelligent artificial system does not have to be emotional. Machines allow for this reasoning.

An essential feature associated to the relationship between the system and its own operation is sometimes lost. Living systems are featured by their autonomy when they operate within their environment. They all are trying to accomplish an essential and broad target of systemic equilibrium, regardless the specific mission. However, the complexity in their behavior is conditioned by their own intelligent abilities.

That is, the systemic equilibrium in living systems—and consequently their wellness, is depending on the level of intelligence of every system. The more intelligent a living system is, the more complex are those principles that define its systemic equilibrium. This reasoning allows for the justification of social emotions in higher beings. Their environments become more complex, because of their higher ability to perceive and exploit knowledge. Consequently, their requirements of wellness and systemic equilibrium are integrating more information and connected parts, and become more elaborated.

So that there is an interesting relationship between intelligence, autonomy, self-adaptiveness and emotion—herein we will refer to reasoning-based adaptiveness, and not to the ability of adapting without exploiting rules. Under this perspective, autonomy might be sight as the condition of self-governing adaptiveness in order to maintain the systemic equilibrium. And emotion as the driver of this autonomy.

1.3 Framework

This thesis has been developed under the framework of the Autonomous System project, a.k.a the *ASys Project*. Short-term objectives of immediate application focus on the design of improved systems, seeking more intelligence in control strategies.

Besides this, long-term objectives underlying *ASys Project* concentrate on high order capabilities such as cognition, awareness and autonomy.

This work is framed within the general fields of *enginery* and *emotion science*, and provides a theoretical foundation for engineering and designing computational emotion for artificial systems. The starting research question that has grounded this thesis aims at addressing the problem of emotion-based autonomy, and this question becomes a general objective. The objective of how to feedback systems with valuable meaning.

Both the starting question and the general objective, have underlaid the study of emotion, the influence on systems behavior, the key foundations that justify this feature in life systems, how emotion is integrated within the normal operation, and how this entire problem of emotion can be explained in artificial systems. By assuming essential differences concerning structure, purpose and operation between life and artificial systems, the essential motivation has been the exploration of what emotion solves in nature to afterwards analyze analogies for man-made systems.

The approach provides a reference model in which a collection of entities, relationships, models, functions and informational artifacts, are all interacting to provide the system with non explicit knowledge under the form of emotion-like relevances. This solution aims to provide a reference model as a framework to design solutions for emotional operation, but related to the real needs of artificial systems.

This thesis has been studied from a wide variety of fields of knowledge. These knowledge categories can be classified into two main area groups: (a) the fields of psychology, cognitive science, neurology and biological sciences in order to obtain understanding concerning the problem of the emotional phenomena, and (b) a large number of computer science branches such as Autonomic Computing (AC), Self-adaptive software, Self–X systems, Model–Integrated Computing (MIC) or the paradigm of models@runtime among others, in order to obtain knowledge about tools for designing each part of the solution.

The final approach has been mainly performed on the basis of the entire acquired knowledge, and described within the fields of Artificial Intelligence, Model-Based Systems (MBS), and additional mathematical formalizations to provide punctual understanding in those cases that have required it. This approach describes a reference model to feedback systems with valuable meaning, allowing for reasoning with regard to (a) the relationship between the environment and the relevance of the effects on the system, and (b) dynamical evaluations concerning the inner situational state of the system as a result of those effects.

1.4 Questions and Objectives

Modern systems are undergoing such a growth in the space of capabilities, that they bring us new problems in complexity management. Fields such as autonomic computing focuses on developing computer systems able of self-managing their own resources, strongly aiming self-adaptation under unexpected events (see Kephart and Chess [123]).

A large number of relevant works in artificial emotion have provide us with essential knowledge related to the comprehension of the problem. However, important questions about shortcomings have emerged from the source of successful approaches. Even when they provide complex solutions that are effectively demonstrated, we still perceive limitations that motivate our feeling of a 'still–life–challenge'.

1. *The question of emotion perception*
   The first essential question is about the perception of emotion. Currently, the major issue is artificial systems do not feel emotion like a conscious phenomena. Consciousness functionality raises quite a fundamental point since it allows the access to the processes' content, sequentiality for consistency in the mental contents, sensory integration to obtain the experience as a whole, meta-representation to obtain abstractions in the form of new usable information, meta-reasoning in order to make meta-inferences about its own inferences, evaluation and learning (see this thesis of Hernández Corbato [106]). This work of Hernández Corbato [106], synopsizes a list of phenomena related to consciousness from a cross-domain perspective: awareness of the world, self(-awareness), introspection, attention and voluntary control. Introspection allows the observation of one's own mental and emotional processes, and these observed processes influence attention, awareness (either be of the world or the self) and voluntary control.
   The question is how this can be essentially solved, in order to provide a first step to solve this challenge.
2. *The question of emotion integration*
   Herein this thesis, cognition is assumed as a high-order ability to intelligently exploit knowledge during an interaction with the environment. Cognitive Psychology studies how humans know the world by means of complex processes that transform sensorial information into usable knowledge (see Neisser [181]). Generally speaking, the term *cognition* refers to mental processes by means of which we manage information, such as perceiving, learning, planning, problem solving, deciding, reasoning, etc. However, this is not always comprehended as an individual and isolated feature. Under the perspective of Militello et al. [168], cognitive activities rarely reside in one individual agent, but instead often happen in the context of collaborative work between different subsystems [168]. And Krevisky and Jordan [130] holds that cognition is the mental action or process of acquiring knowledge and understanding through thought, experience, and the senses.
   The question is how emotion can be integrated in order to provide means to improve the exploitation of models.
3. *The question of uncertainty management*
   The matter herein involved is how life beings are capable of managing themselves within the environment, even when they do not have complete information about it. Unquestionably, any unknown or unstructured environment cannot be modeled

on its entire form at design time, because models always depend on observation. However, since we are the designers of the system, we are in possession of the entire model of this system and this gives us a new course of action to use in the design of control strategies. Since we cannot define all the external environment that is having effects on the system, the solution is to decrease the influence of the external environment over the system, at the same time that we increase the influence of its internal state.

We can explain this idea on the basis of a normal control operation. Under the vision of Astrom and Murray [9], feedback loops are the tool by which the system senses the environment, compute control strategies and execute actions to ensure stable dynamics (i.e. bounded disturbances have to induce bounded errors) [9]. Disturbances is a key facet of control design and they are modeled from the most simple forms (as a deterministic signal or random signals) to the most complex models of uncertainty. Nevertheless, the design of accurate models of uncertainty is not so easy and, as a consequence, system predictions made on the basis of these models might not be effective enough resulting in unsuccessful control strategies. Our objective is to focus the control strategy on the consequences of the disturbances rather than on the disturbances themselves. If we were to have an additional control system whose goal would be to ensure the control of the desired behavior of these disturbances (i.e. responsiveness under changes, attenuation, etc.), we would be increasing the capability of the system for controlling these uncertainties. In the end, systems might be able to monitor and measure the inner effects of these disturbances.

This thesis has been planned aiming to study the broad objective of emotion-based autonomy. That is, the improvement of systems' self-evaluation abilities, in order to better exploit knowledge, self-adaptiveness and autonomy.

To this end, the objectives of this thesis can be described under the following goals:

1. *The analysis of emotion*
   It refers to the understanding of the purpose of emotion in living systems, in order to provide the artificial systems with analogous characteristics.
2. *The scrutiny of models*
   It focuses on the need to understand the essential concepts that defined each theory with respect to the tool that might convey analogous functionalities in artificial systems.
3. *The principles for computational emotion*
   This is an essential objective of this thesis. There are required principles under which to design the artificial approach. However, we argue that those principles should be supported by a full range of common features that comes from different theories. This purpose is motivated by the objective of a general explanation focusing on artificial systems. It is essential for the purposes of this thesis the identification of principles from the basis of accepted theories in emotion science, in order to find foundations that might be placed appropriately to conform a computational theory.

4. *The analysis of computational emotion*
   Principles for computational emotion need to solve the essential questions of (a) the nature of an artificial emotional goal, (b) the emotional space of an artificial system, (c) how the emotional assessment is required in order to influence the system, (d) how the artificial system might obtain perception of emotional value, (d) how the result of the emotional operation can be conveyed to the system cognition and behavior, and (d) how the system might obtain some type of emotional representation.
5. *The integration of emotion science theories*
   One essential transversal objective in this thesis, is the goal of providing a reference framework under which every emotional theory can be used. Once implemented the system, the principles of emotional objectives and appraisals will be determined by any chosen emotional theory.

Ultimately, it is important to present some final remarks to highlight the essential connections of this research with some of the general problems in enginery science.

1.5 Dissertation Structure

Following this introduction, the following chapters are forming steps as guidelines in order to build the final hypothesis. They are structured this way to provide the required background to every idea herein discussed, proposed or simply used. Chapters 4, 5 and 6 are the main body that provides the understanding about what is proposed in this work.

The path through which this hypothesis has been built, has required a profound study in order to obtain the guidelines over which to obtain the principles of emotion design. Thus, Chap. 4 is aimed to provide the main rationales of this thesis that thereafter conform the basis of the proposed model. Chapter 5 describes the model on the basis of these rationales, and Chap. 6 provides a discussion about the proposed model and its relation with the state of the art. The convention in this thesis is to use of the symbol ♣ to mark (a) every proposed idea that is an original contribution of this work, or (b) every referenced idea that will be accepted under the principles of design of this current work.

There are some rationales in Chap. 4 that are entirely based on other works. These rationales reference every specific theory or theories to which they refer, and the rationale will refer to the acceptance of these theories for the purposes of this work.

1. *Introduction:*
   This chapter provides the essential knowledge about the antecedents of this thesis, and offers a portrait of the scope, background, framework, research questions and goals of this work.
2. *Models:*
   It is focused on the binding of models in order to determine the relationship

between them, concerning foundations and solutions. It intends to provide the perspective under which models are used within the scope of this thesis.

3. *State of the Art:*
   This chapter intends the particular purpose of understanding emotion science from the wide range of perspectives that can be found in the state of the art. It provides a general overview on emotion science, trends in biological emotion research, general directions in which artificial emotion research is being studied, analogous questions about emotion modeling, and a final discussion.
4. *Principles for Computational Emotion:*
   This chapter is the theoretical cornerstone of this thesis. There are discussed assumptions and theoretical strategies underpinning the emotion science. Common theories are analyzed and interpreted from a computational perspective in order to describe and interpret emotion from the point of view of an artificial system. This chapter proposes 61 thesis-rationale that will support, and will be related to every modeling decision in the following chapter. They are denoted as *rationale* because they are the set of reasons and logical basis proposed in this thesis, for the course of the subsequent modeling decisions.
5. *Computational Emotion:*
   This chapter provides an exhaustive description of the proposal, a simplified description of an emotional system to assist the performance of computational emotion, and the subsequent emergence of artificial emotion. The description of the reference model is made on the basis of the principles expressed in the previous chapter under the form of thesis-rationales. The main purpose of this chapter is to provide a useful description and interpretation of how computational emotion might work, and the explanation of the different foundations of the phenomenon. These descriptions focus on a later subsequent implementation.
6. *Discussion:*
   This chapter aims to provide a detailed treatment between the particular solution of this thesis, and the current challenges in enginery science. Here, the correlation between the proposed approach, and what it might be expected from an emotional model is firstly analyzed. Later, we connect the challenges in autonomy and self-adaptive systems with the solution proposed. There are additionally provided essential descriptions of the nature, scope and meaning of artificial emotion under the understanding of this investigation.
7. *Future Works:*
   Future works are focused onto two essential lines: the one related to the implementation of the current proposal, and those related to the improvement of system features. There are (a) described those studies that will provide improvements in these two referred lines, (b) showed conceptual lines that need for a deeply study, and (c) delineated those questions that require of an additional research in order to provide the complete functionality.

1.6 Assumptions, Limitations, and Scope

This research falls within the broad field of emotion science and it takes the engineering perspective. We investigate fundamental processes from biology to build computer models that implement and explain these processes. The references in this thesis are works in the state of the art of emotion science, and they constitute the baselines and principles under which this thesis has been developed.

Certain features of life systems are still without a complete explanation, and this research is completely conditioned by this issue. The scientific community has studied and analyzed different perspectives of the emotional phenomenon isolating effects and integrating subjectivity within theories commonly under observation.

It has been aimed the discovering of connections between those relevant theories that have explained essential parts or elements of the emotional phenomenon. And among the broad range of theories, we have chosen those with explanations that allow for computational models. This thesis proposes a single theory which is composed of several well-coordinated theories. The discovering of connections between those relevant theories that have explained essential parts or elements of the emotional phenomenon have deserved special attention.

It is strongly highlighted that the emotional model provided in this thesis is not argued as the correlation of the real phenomenon. The reason is that this is just only one way of understanding how artificial systems might be provided with analogous functionalities to those of emotional living systems in order to improve their functionality.

Chapter 2
Models

Here are some descriptions of *model*, according to distinct standards and norms in computer science:

> Model is an approximation, representation, or idealization of selected aspects of the structure, behavior, operation, or other characteristics of a real-world process, concept, or system
>
> [Software Engineering (IEEE 610.12-1990)]
>
> Model is a representation of a real world process, device or concept
>
> [IEEE Standard 1233–1998 (R2002)]
>
> Model is a representation of something that suppresses certain aspects of the model subject
>
> [IEEE Standard 1320.2–1998 (R2004)]
>
> Model is a related collection of instances of meta-objects, representing (describing or prescribing) an information system, or parts thereof, such as a software product.
>
> [IT framework (ISO/IEC 15474–1:2002)]
>
> ...a semantically closed abstraction of a system or a complete description of a system from a particular perspective.
>
> [IT framework (ISO/IEC 15474–1:2002)]

From the viewpoint of Evans [74], the ingredients of an effective modeling are: (a) binding the model and the implementation, (b) cultivating a language based on the model, (c) developing a knowledge-rich model, (d) distilling the model, and finally (e) brainstorming and experimenting.

This chapter focuses on the first principle: the binding of models in order to determine the relationship between them, concerning foundations and solutions.

M.G. Sánchez-Escribano, *Engineering Computational Emotion—A Reference Model for Emotion in Artificial Systems*, Cognitive Systems Monographs 33,
DOI 10.1007/978-3-319-59430-9_2

2.1 Introduction

This thesis essentially aims at building *models*, but *model* is a broad concept whose meaning is not so clear as it might seem. It is thus necessary to clarify what the concept of *model* refers to in this work, in order to make this work more clearly comprehensible.

There exist a vast number of purposes towards which models can be used, e.g. in point predictive financial, meteorological or economics models, decision making and behavior models, data base and data structure models, software architecture models, network models, mathematical models, etc…The concept can also be understood from limitless perspectives such as patterns of states, patterns of data, relationships, arrangements, graphs, sequences, blueprints, etc…as well as in an endless set of forms and details.

One generalized view of *model* is the one that comes from the work by Rosen [222]. He argues that two systems (A) and (B) are in a modeling relationship one to each other, if it is the case that entailments in the model A can be mapped to entailments in the model B. Basically, this seems to be as much as is required in order to obtain a general idea of *model*. However, the construction of entailments as well as the task of chosing these entailments and their final forms, is not a simple problem. And the domain where the model is built imposes constraints on these final decisions.

By fixing this work in the specific fields of *cognitive science*, one becomes fully aware of the complexity of human-like processes and the consequent difficulties in modeling tasks. This has caused a collection of models such as *agents*, *software architectures* and additional computer science based artifacts that are usually found together or resemble one another in common solutions. Additionally to this, *cognitive science* conforms a multidisciplinary field in which computer science, artificial intelligence, biology, psychology, philosophy, mathematics, and several other fields, are part of a larger amount of theories to which a lot of scientist contribute. And this results in the laxity not only regarding the key concepts that ground the theories, but also regarding the models used in the final solutions.

This thesis needs for a particular purpose successful descriptions, regarding both the problem as well as the final solution. It is beyond the bound of models comprehension that the problem could approach an understandable result. So that, the concept of *model* needs to be putted in context to describe the circumstances that surround this thesis.

2.2 Agents

Generally speaking, models can be conceived as abstractions of the reality.[1] According to the perspective of Booch et al. [26], *models* are abstractions that provide a simplification of reality avoiding those details that are not relevant [26]. Given the

[1]Formally defined as abstractions in ISO/IEC 15474–1:2002.

broad use of models in *Cognitive Science* and *Artificial Intelligence* and for the explanatory purposes of this chapter, I will assume the vision adopted in the work by Russell and Norvig [223] regarding models.

The first question about models is the one regarding *agents*: the conceptualization, definition and use of this type of models. A large number of ideas, discussions and dissertations are shown in the state of the art, resulting in closely connected arguments and appropriate definitions depending on the matter at hand. There is a relevant survey conducted by Nwana [188] regarding *agents*, arguing that they "tend to use the word slightly more carefully and selectively":

> When we really have to, we define an agent as referring to a component of software and/or hardware which is capable of acting exactingly in order to accomplish tasks on behalf of its user. Given a choice, we would rather say it is an umbrella term, meta-term or class, which covers a range of other more specific agent types, and then go on to list and define what these other agent types are. This way, we reduce the chances of getting into the usual prolonged philosophical and sterile arguments which usually preceed the former definition, when any old software is conceivably recastable as agent-based software.
>
> Nwana [188]

This work classifies *agents* regarding their abilities: (a) the ability of movement classifies agents into *static* and *mobile*, (b) the ability of thinking-like separates them into *reactive* and *deliverative*, and (c) some essential ideal and primary attributes result in the differentiation between *autonomous*, *learning* and *cooperative* agents (Fig. 2.1).

Nwana [188] discusses and expresses some judgements about the proposed classifications and characterizations made in their work, and shows a relevant discussion about what is and what is not an agent:

> In general, we have already noted that a software component which does not fall in one of the intersecting areas of *does not count as an agent*
>
> Nwana [188]

Even without addressing *agency* characterization (in the sense argued by Minsky [170]), they hold that "modules in distributed computing applications do not constitute agents either as Huhns et al. [112] explain". The argument is that *agents*

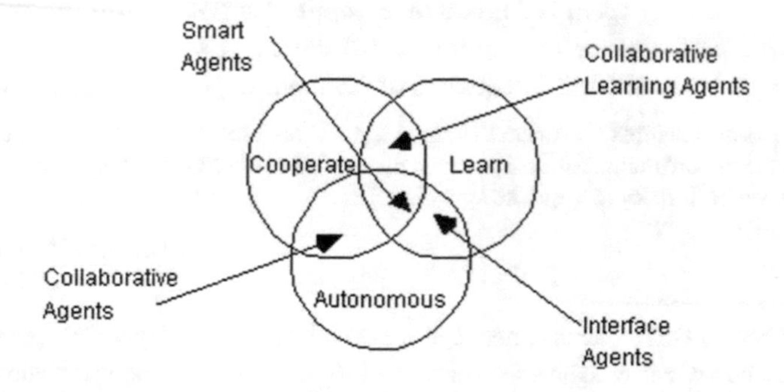

Fig. 2.1 Minimal characteristics to derive four types of agents according to Nwana [188]

Fig. 2.2 Classification of Software Agents according to Nwana [188]

should be (a) more robust and *smarter* than modules in distributed computing, (b) should involve high-level communication (in contrast with lower-level messages in distributed computing), and (c) should operate at the knowledge level (see Newell [184]) in contrast to distributed computing that operates at the symbol level Fig. 2.2.

Russell and Norvig [223] describe the viewpoint of a "rational agent approach" able to act in a rational manner:

> An agent is something that acts ('agent' comes from the Latin 'agere', to do). Of course, all computer programs do something, but computer agents are expected to do more: operate autonomously, perceive their environment, persist over a prolonged time period, adapt to change, and create and pursue goals. A 'rational agent' is one that acts so as to achieve the best outcome or, when there is uncertainty, the best expected outcome.[2]
>
> Russell and Norvig [223]

Russell and Norvig [223] identify the concept of *rational agent* as fundamental to the approach to artificial intelligence [223], and consider a more concrete concept of *intelligent agent* that always maintains an interaction with the environment by means of sensors and actuators (Fig. 2.3). Each agent has an *agent function* that maps any given percept sequence to an action, assumed as an abstract mathematical description that maps sensors and actuators and that needs for the entire percept history. However, this work argues an essential difference with the concept of *agent program* required in artificial agents, a concrete implementation that is running within some physical system.

So, Russell and Norvig [223] assume that *rational agents* accomplish some type of performance measure in order to: (a) interact with their environment, and (b) assess consequences that concern each change caused on this environment. With this assumption, a *rational agent* is defined by accepting that *rationality* depends on four things: (a) the performance measurement, (b) the agent's prior knowledge, (c) the actions knowledge, and (d) the sequence of perceptions till the current time:

> For each possible percept sequence, a rational agent should select an action that is expected to maximize its performance measure, given the evidence provided by the percept sequence and whatever built-in knowledge the agent has.[3]
>
> Russell and Norvig [223]

---

[2]Russell and Norvig [223] refer to correct inferences under the 'laws of thought' approach to AI.

[3]It is essential however to enlighten the notice made by authors regarding that "rationality is not the same as perfection", since "perception maximizes actual performance" (argued as not possible in reality) while it is hold that "rationality maximizes expected performance".

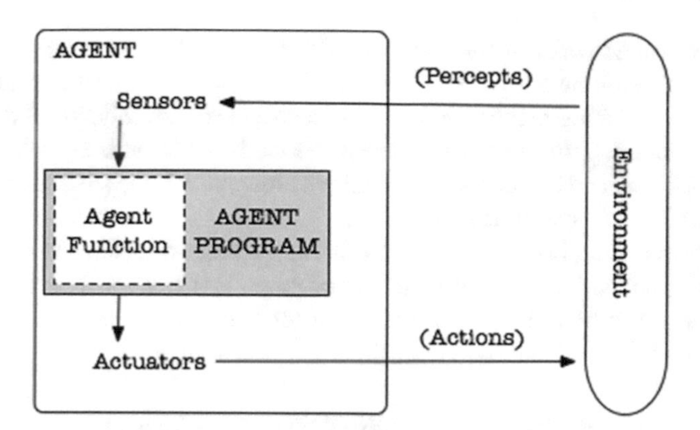

Fig. 2.3 Conceptualization of an artificial agent (adaptation from the viewpoint of Russell and Norvig [223])

So far, the arrangement of an *artificial agent* (either rational or not) depends on the *agent program*. That is, a program that performs a successful *agent function* to map percepts and actions. Depending on the features of this program, the agent will deploy a different range of behaviors, thereby Russell and Norvig [223] classify *agents* on the basis of their rationale degree into: (a) *simple reflex* agents, (b) *model-based* reflex agents, (c) *goal-based* agents, (e) *utility-based* agents and (f) *learning* agents Russell and Norvig [223].

The *simple reflex* agent implements a set of " If-Then" rules to provide actions in accordance with the current state of the agent. From this starting point ahead, there are essential differences regarding the growing of capabilities along the other four types of agents. A '*model-based reflex agent*' is conceived as an agent capable of keeping track of part of the world that it cannot see at the current time. It depends on the percept history, and thereby reflects aspects of the current state that actually are not being perceived. The agent uses a "model of the world" in order to update its inner state, knowledge about (a) how the world evolves with no agent's interaction, and (b) how the agent actions affect the world. Additionally, *goal-based* agents define descriptions about situations that are desirable, thereby they perform more complex tasks such as searching and planning. The conceptualization of an *utility-based* is a step up for the rationale ability by means of internalizations of the *performance measurement*[4] (see the book of Russell and Norvig [223] for additional details).

One essential objective beyond the field that bounds this thesis, is to obtain understandable computer models on the basis of knowledge that comes from fields that are not technical at all. The requirement of formalizations is however an essential need for the particular purpose of implementing computer programs, and these programs are commonly related to *intelligent agents* that approach different cognitive abilities. This way, the *agent program* results in a new concept of *software architecture* that helps agents to deploy cognitive capabilities.

[4]Russell and Norvig [223] define this concept as the score related to a concrete sequence of environment-states.

Software architectures are viewed by Garlan [86] as a "key milestone in an industrial software development process" that simplifies our ability to understand large systems by introducing higher levels of abstraction [86]. Architectural descriptions provide partial blueprints where there are integrated components as well as relations and dependencies, and can expose dimensions along which the system might evolve allowing for higher capabilities [86].

The concept of *software architecture* is thus a central artifact to the nature of *cognitive science* and consequently, to the problem that this thesis is handling. A a result of this discussion, I will assume hereafter that our system is an agent in the form argued by Russell and Norvig [223]:

$$Agent = Architecture + Program_{(Agent)}$$

And the general notion of $Program_{(Agent)}$ will be conceived under the concept of a *software architecture*. However, I will enlighten the issue that *software architecture* herein is not the same concept of architecture as used by Russell and Norvig [223]—since they refer to the body of the agent, either physical or any type of computer program.

Consequently, since the objective in this thesis focuses rather on the *agent program* than in the shell of the agent, I will redefine the concept of agent under the following terms in order to reduce terminological laxity:

$$Agent = (\text{Body Architecture})_{(Agent)} + (\text{Software Architecture})_{(Agent)}$$

I will however use interchangeably the concepts of *system* and *agent* along the dissertation. That is, I will conceive this agent as a generic system that will perform emotion by means of a (Software Architecture)$_{(Agent)}$, with the requirement that (Body Architecture)$_{(Agent)}$ will not constraint the final approach. And that they both follow such descriptions as to perform with success regarding the environment perception (i.e. sensors) and action (i.e. actuators).

Thus, I will refer to our *Agent* as *AGSys (Agent Generic System)*. And following the argument of Russell and Norvig [223], I will assume the architecture either as a physical platform or a computer program with no allusions related to this concrete feature of *AGSys*.

2.3 The Emotional Agent

An essential feature in this work is that which refers to the software architecture of *AGSys*: I will assume that an emotional functionality will be conceived as a transversal attribute concerning the AGSys program and architecture. This way, emotion might be added to any particular agent that has some sort of rationality. I am not meaning herein that the entire software architecture of *AGSys* will be transversal, but only the part that regards the emotional solution. This section will describe the main foundations and principles regarding models, in order to build a computational solution of emotion.

Because we pay more attention to our own perception than to specific formalizations, our common sense moves us towards the idea that *emotion* is closely related to the concepts of *cognition* or even *self*.

As previously argued, *AGSys* can be considered as an *agent* which is coupled with the environment [223]. Besides this, *cognition* is commonly viewed as a process of acquiring knowledge through senses, experience and thought. Consequently, I will assume those principles and foundations on topo of which our research framework is situated. Thus, I will assume the task of modeling cognitive processes by holding the principle of *Model-Based cognition* argued by [230]:

> Model-based cognition: *A system is said to be cognitive if it exploits models of other systems in their interaction with them.*
>
> (see Principle 1 in [230])

So, what might be expected is that the (Software Architecture)$_{(Agent)}$ might endow *AGSys* with essential attributes that allow it to perform some type of cognitive abilities. And I will assume that this (Software Architecture)$_{(Agent)}$ will be designed according to the principles argued by Sanz et al. [230].

Within the fields of artificial intelligence, cognitive science and robotics, *cognition* somehow concerns intelligent behavior and interchange of information with the environment. The Artificial Intelligence field is focusing more and more on improving intelligent systems. And it inspires the solutions in nature, in order to construct systems that perform at high levels of reasoning. The *cognition* feature in artificial systems should—at least—involve the essential features that characterize the *model-based*, *goal-based*, *utility-based* and *learning* agents conceived by Russell and Norvig [223].

To this end, (Software Architecture)$_{(Agent)}$ becomes a complex architecture in order to provide the grounds for high level reasoning. They are software architectures that are terminologically known as *cognitive architectures*. That is, models that devise a representation of the human reasoning [223], and that nowadays represent a relevant approach towards intelligence and cognition in artificial systems.

Getting back to the vision of Russell and Norvig [223], goal information establishes circumstances that are desirable for the agent, and that allow for rational thinking concerning a set of possible paths. As argued by the authors, an agent is *happy* or *unhappy* regarding the goal achievement degree. As the authors refer to *utility* instead of *happines* (as “the quality of being useful”), we can think of an agent able to measure the degree of success concerning usefulness. This is argued as an agent feature; the agent implements an *utility function* in order to recognize the desirability of choosing each possible path.

By conducting a deeper analysis from the viewpoint of emotion science, there are some aspects of this characterization of *happiness* that require some additional discussion. To begin with, I completely agree with the characterization of *happiness* made by the authors. What I hold is that a more complex concept of *happiness*, might endow systems with a more complex ability to reason about the desirability of actions.

Happiness seems to be related to some type of *system wellness* regarding its *inner situational state*, as well as related to the *situational state of the environment*. The flexibility and the adaptiveness of the agent might be related to the capability of managing its *wellness*, rather than to the capability of choosing among a wide range of complex paths and assessments.

Additionally—and even if it seems obvious, an agent cannot realize anything that it has not been prepared to perceive. Environment involves interaction with uncertainty, and uncertainty is not devised. Therefore, the system is not able to be directly relied on to act concerning unknown occurrences Fig. 2.4.

Herein, I am not talking about those occurrences that an agent might not be able to perceive because of sensor constraints. I am alluding uncertainties related to lacks of knowledge due to which the agent cannot make inferences, nor trigger actions, i.e. a scenario in which the agent cannot properly reason since it has not enough knowledge to judge and/or act.

Emotion might allow for new scenarios into the agent in the form of new grounds for wellness assessment, and this might provide a new inner scenario associated to references for wellness inside of the system.

Uncertainty cannot be directly managed, but we improve the probability of the agent to make an optimal reasoning under environmental uncertainty. Thinking about

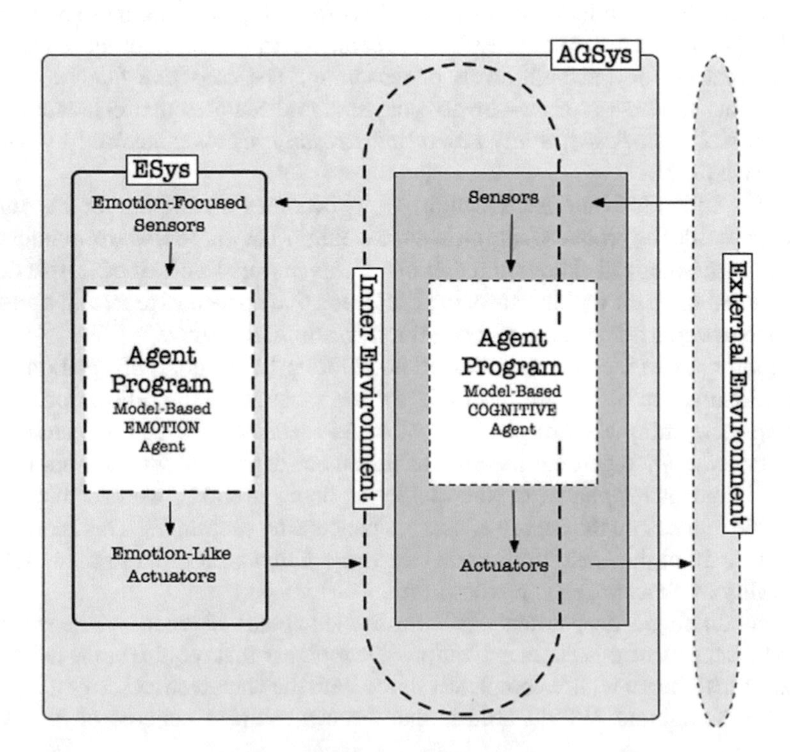

Fig. 2.4 Structure of an Emotional cognitive agent under the approach proposed in this thesis

the design of models of knowledge, the range of environmental events that might cause an effect (either positive or negative) on the agent is unmanageable. Consequently, the objective of optimal responses becomes a goal with low degree of success.

Thinking of life agents, the same emotion can emerge from different sources of interaction with the environment. Somehow, each emotional reference allows for measuring a range of environmental events, and this same reasoning can be argued in artificial agents. Inner references might allow for controlling ranges of uncertainty by means of the effects that they cause on the agent, rather than directly attend the external event.

By using the conceptualization of Russell and Norvig [223], I hold that the *utility function* might be improved by means of emotional attributes. With a simple description, it can be assumed as a high-order *utility function* (essentially an internalization of the *wellness* performance measure), that regards the relationship between the *utility function* and the situational state of the agent.

This idea is conceived by decoupling the external and the inner environments of the agent (i.e. *AGSys*). To this end, I will argue the requirement of an—additional—inner agent with the same conceptualization of agent assumed till now. This new agent that I will name *ESys (Emotional System)*, will be embodied into *AGSys*. And consequently, it will be characterized as living within the inner environment of *AGSys* as illustrated in Fig. 2.5.

This way, the *utility* (conceived as an attribute of the agent), the *performance measurement* as well as the *utility function* of *ESys*, will be related to sequences of inner situational states of *GSys* (instead of sequences related with its external environment).

And this is the proposed scenario in this thesis: a model of reasoning under the reference of *wellness*.

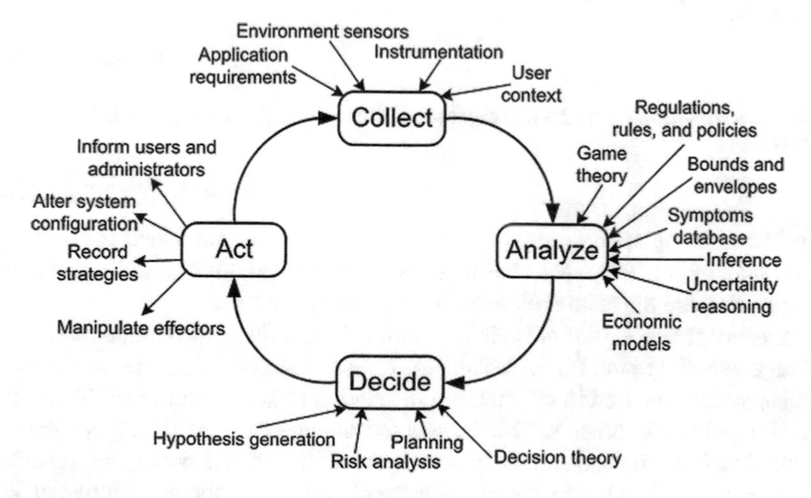

Fig. 2.5 Autonomic control loop. (*Source* Brun et al. [37])

2.4 Technological Portrait

> Model-Based Design is emerging as a solution to bridge the gap between computational capabilities that are available but that we are yet unable to exploit. Using a computational approach in the design itself allows raising the level of abstraction of the system specification at which novel and differentiating functionalities are captured.
>
> Nicolescu and Mosterman [186]

The purpose of this work by Nicolescu and Mosterman [186] is to provide a comprehensible portrait regarding *Model-Based Design (MBD)* focusing on *embedded systems*, that is, "systems designed to perform a dedicated function, typically with tight real-time constraints, limited dimensions, and low cost and low-power requirements" [186]. Even when this thesis does not define those constraints that *embedded systems* regard, the arguments hold in this book are those that cognitive science needs in order to obtain higher levels of abstraction.

This thesis will assume the task of modeling cognitive processes by holding the principle of *Model-Based cognition* argued by Sanz et al. [230]. It is essential however to clarify the viewpoint from which this principle is understood in this thesis: I will assume that *Model-Based* is not only referring to knowledge based models. I will understand the concept of model as involving a wider meaning, and referring to several other types such as agents or software architectures.

There are several branches in computer science within which models are used to obtain different solutions. They are all distinct parts of a broad field, in which the activity focuses on improving solutions by growing the level of abstraction.

Greenfield et al. [95] describe the field of *Model Driven Design (MDD)* as the one that embodies ideas and technologies that predate object orientation [95]:

> the confluence of component based development, model driven development and software product lines forms an approach to application development based on the concept of software factories.
>
> Greenfield et al. [95]

> Factories are software artifacts that encapsulate the knowledge needed to create a complex object
>
> Evans [74]

In software engineering, models have become a well-established form of information integration. There are a large number of applications and fields of research where models are the central element to approach solutions.

The concept of *adaptation* is closely related to *emotion*, since emotion is considered an essential ground for adaptiveness. As it happens in life systems, the idea of adaptive systems is used in engineering in relation to how systems might optimally deal with their environments. That is, how to maintain their resilience by improving their feedback mechanisms. This becomes the ability of self-managing at runtime, to manage a situational state caused by some change. From the perspective of Vogel and Giese [264] this is a question of how the system configures and reconfigures itself, continually tuning and optimizing, protecting and recovering while keeping its complexity hidden from the user [4, 264].

This topic has been studied in a large number of areas. And it searches for methods in order to encompass the influence of change wherever it comes from, this is one of the main issues of software engineering for self-adaptive systems (see Cheng et al. [45] for additional details about self-adaptive systems roadmap).

Autonomic Computing (AC) (see Kephart and Chess [124]) is an example of the new paradigms addressing those exigent needs from the viewpoint of self-adapting software. It deals with the self-management ability of distributed resources to attain system-level goals. Self-adaptive software aims to adjust different artifacts or attributes in response to changes in the self and in the context of a software system. The domain of *Self-X systems* (see Sanz et al. [229] and Babaoglu et al. [12]) addresses the architectural aspects of systems that are autonomous, in the management of certain architectural traits.

The area of *Model-Integrated Computing (MIC)* [255] studies how to extend the scope and use of models to maintain a correlated sequence of changes between the model itself and the complex system in which it is implicitly or explicitly embedded. A main feature of MIC is that it tries to match modeling paradigms to the needs of the systems engineering domain.

As previously mentioned, *Model Driven Software Development (MDSD)* [95, 245] focuses on using software models for increasing the quality of software and improving the effectiveness of the software process. A more open-domain line of research is the one is globally referred to as Model Driven Engineering (MDE) [122]. *Model Driven Engineering (MDE)* is a software methodology focused on the exploitation of models. Molina [174] denotes *Model Driven Engineering (MDE)* as a modern discipline situated within the software engineering, that aims at the systematic use of models in order to improve productivity, maintainability and interoperability in software systems [174]. This work describes a number of branches within the paradigm of *MDE*, focused on addressing three main types of solutions in software engineering: (a) forward engineering for creating new applications (*Model Driven Design (MDD)*), (b) reengineering, and (c) *Models@runtime* for adaptive systems.

Models@runtime (see Fouquet et al. [76], Blair et al. [24] and Daubert et al. [58]) creates models used to reason about the system and its operating environment at run time. And this field is directly related to the challenge of dynamic identification of requirements, with the goal of ensuring the reliability of systems when unplanned or unexpected events (i.e. changes) occur [45].

This thesis addresses the issue from the analysis of the cognitive architecture of biological emotion. There are plenty of examples of cognitive architectures that address partial aspects of human cognition (see the following chapter regarding state of the art). From a more systems centric view, it goes from studies about end-means patterns as hierarchical organizations for structuring goals and means, to a large number of applicable analysis about complexity in engineering systems (see Ashby [6–8] for self-organizing systems, and Axelrod et al. [11] for harnessing engineering), organized systems [159], risk management [224] or entropy growth management (see Haghnevis and Askin [99] for adjustment of structural artifacts to respond to the challenges of the environment).

The focus on emotion is not on the perceivable aspects (see the theory of Ekman [66]), but on the structural traits concerning the adaptation of system structure to shape behavior [231]. This thesis will focus on runtime self-awareness paradigms of relevance for this paper such as self-adaptiveness, model-centric systems and cognitive architectures

As previously introduced, this thesis proposes to decouple the external and the inner environments of the agent (i.e. *AGSys*). Consequently, it is required an inner agent with the same conceptualization of agent assumed till now. And this new agent (i.e. *ESys (Emotional System)*) will be embodied into *AGSys* (see Fig. 2.5). This regards the field of *agent-based modeling*, a relatively new approach "to modeling systems composed of autonomous, interacting agents" (see Macal and North [154]).The paradigm *agent-based modeling* understands the modeling of complex systems by means of agents with their own rules, interacting one to each other (see John [119] and Kauffman [121]). It allows to observe the effects of the diversity that exists among agents (in attributes and behaviors), as well as to provide means by which to model the dynamics of complex systems and complex adaptive systems.

The complexity of an emotional agent requires the availability of models to attend the key issues related to the changes in its dynamics, as well as an interconnected design that might address the exploitation of the functional structure, the organizational structure, and the knowledge base of the agent.

This thesis argues the requirement of *agent-based modeling* in order to attend essential requirements of dynamics related to internal interactions, an essential and critical feature in emotion science. And the requirement of additional paradigms such as *models@runtime*, *self-awareness*, and paradigms of relevance such as *self-adaptiveness*, *model-centric systems* or *cognitive architectures* in order to obtain particular solutions, assemble artifacts or create blueprints for interaction.

There is no doubt about the need of models to incorporate usable knowledge in cognitive systems. Models are abstractions of reality that contain just the essential aspects of this reality concerning the system [238]. The function of a model refers both to the interpretation and the understanding of the system, as well as the drawing of conclusions in the form of other subsequent usable models. Under the approach of *Model-Based design*, the system is able to take part in its own control strategy—an indispensable feature in emotion science.

2.5 On Appraisal

When we refer to *artificial emotion*, the feature *artificial* intrinsically implies the requirement of some type of *computable process* (with no allusion by now to the nature of this process). I will hold a *carrier process* and the final emotion as two essentially different computational artifacts. This *carrier process* is conceived as the series of steps taken in order to endow the system with the ability of deploying emotion. I will thus refer to the *carrier process* as the one whose operation will ground the complete emotional phenomena, with no allusion to any associated process that might help to conform this phenomena.

If final emotion is required, there is need for a particular finite algorithm that computes a concrete emotion with any precision degree. Consequently, I conceive this *carrier process* as a finite process, algorithm, function or so, able to compute an emotion with any degree of precision.

One of the major paradigmatic contributors in the field of artificial intelligence is Turing [259], a fundamental reference for a large number of works in order to obtain measurements on the degree of intelligence in artificial systems. Clearly, Turing,[260] proposed a set of foundations and principles for intelligence that are widely known and accepted (principles that were wrapped under the *Turing's test* [260]):

> (...) all the skills needed for the Turing test also allow an agent to act rationally
>
> Russell and Norvig [223]

However, there are several other works that are quite relevant to the field of artificial intelligence and that regard the *computability*. Turing [259] conceived the notion of *computational* number as the one that can be computed by means of a finite process:

> The computable numbers may be described briefly as the real numbers whose expressions as a decimal are calculable by finite means
>
> Turing [259]

Therefore, there is a comprehensible rationale in thinking of a *carrier process*—aimed to build emotion, characterized with the distinctive attribute of being *finite*.

Initially, this work by Turing [259] referred to the *real numbers* as the artifact to be computationally described, while we are herein referring to the *emotion*. Somehow, the nature of *real numbers* involves the feature of *continuity* [271], and this is a shared feature with the process of *emotion* (see [71]). And the reasoning of Turing [259] is approaching a solution that relates *continuity* and *computability*: the capability of computing a continuous artifact. Since it has been widely solved when we refer to the normal work of our computers, it is relevant to make an analogous analysis of computability when we refer to computational emotion in order to understand the form of the *carrier process*.

The nature of emotion is essentially more intricate than the real numbers, and conforms a set of complex processes which involve a large amount of elements working and interacting all together. However, the same reasoning made by Turing [259] with respect to real numbers can be arguably applied regarding emotion. This reasoning refers the fact that, as it appears, emotion (as a continuous artifact) might become a computable emotion just in those instances when the process that triggers this emotion can be *finite* by nature.

I am searching for the capability of computing a phenomena that is continuous in its real form (i.e. a continuous appraisal is progressively conforming an emotion). However, the *carrier process* responsible of converting a *continuous real phenomena* of emotion into a *computable artificial emotion* needs to be finite and should allow for computing this artificial emotion with any required degree of precision. This way, a concrete emotion might be computed regarding the feature of a continuous artificial emotion, even allowing a new parameter of "precision gain" (which might allow for new inferences). It seems as if this *carrier process* might be the *appraisal process*. Consequently, this problem of artificial emotion concerns the searching for analogies regarding the *computing machines* of Turing [259].

Saying this in a different way, the research question is: how should we build a *computing appraisal machine* so that we obtain assessments that help construct an artificial emotion. The answer to this question summarizes in determining which are the "scanned symbols" (of which the appraisal system is being "directly aware" of), which the "states (q)", and which the "configurations" that might help obtain assessments to determine a concrete emotion.

It is widely accepted that emotion works to achieve the stability in the state of the goals of the system. Somehow, emotion interrelates processes in order to maintain a systemic equilibrium of the system. I argue that those inner goals related to the emotional processes as well as their stability, might be the artifact by means of which the *appraisal process* can be built. The logic I follow to make this inference is that, it seems to be clear that emotion emerges from the source of a systemic unstableness related to these inner goals. So that, unstableness (regarding inner goals) drives the behavior of the system in order to recover stability.

2.6 The Exploitation of Models

There is no doubt in cognitive science about the need of models in order to integrate knowledge. As has been described, models are abstractions of the reality that contain just the essential aspects of this reality that concerns the problem domain [238]. The function of models regards essential functional objectives for artificial systems. It is thanks to modeling that systems might be able to be easily modified to respond to altered circumstances or conditions. And it is through modeling that they obtain implicit or explicit meaning.

Clearly, there is no debate about the topic of models and its relevance. The problem is to analyze which aspects of the emotion reality concern the system (according to the principles of [238]), in order to allow the exploitation of these models for the system's advantage. In the end, models will affect how successful a system will be. And this question of aspects-searching is the one this section focuses on.

I have analyzed some essential requirements concerning the models that might help to cope with the process of appraisal and the dynamics of the system. However, there is a transversal issue for discussion which concerns the exploitation of models.

Even when the extent of emotion will be covered in the following chapters, I will here clarify a few essential concepts in order to ground some arguments. From the perspective of cognitive science, emotion (and specifically the notion of *appraisal*) is conceived as a key reasoning component for human-like intelligence [94]. The idea of *appraisal* regards the notion of some type of dimensions that ground the emotional assessment [71, 148]. The inner goals of the system and the *appraisal* seem to be intrinsically related, and the stability of these goals will influence the *appraisal state*.

Appraisal assessment endows systems with the notion of *value related to something*. This is intrinsically tied to the perceptual capability of the system in order to obtain this *value*: In my view, the capability of reasoning is responsible for building the *relation to something* (with no allusion about the process through which this relation is built).

The notion of *action readiness* is the concept that represents those essential processes triggered when required, that is, on punctual states of the system [79, 202]. There are broadly assumed *theories of readiness* (see [80, 81]) that argue meaning from the basis of these readiness processes. That is a critical issue sometimes referred to as part of the experience of what happens (see [54]). This notion is also related to the perceptual capability of the system.

Clearly, the system should be capable of exploiting models in order to understand the particular patterns of functions that are running at a given time. The reason is that comprehending the patterns of these functions might provide further information with respect to the appraisal evaluation. So that as crucial as the models themselves are, the building of artificial emotions requires the engineering of an essential *emotion experience*[228].

As argued in the relevant work of Hayek [105], the human being interprets in the light of experience. This results in the idea that any model that is not being experienced cannot be fully understood, and that those models that still have not been exploited by the agent are not complete models. Hayek [105] describes an example in order to support his argument: The experience of a mirror image is an exploitable model concerning the causal relationship between the visual system of the agent, and the functionalities that concern its visual perception.

The significance that I perceive from the argument of Hayek [105] is that, even when the model might provide an artifact as the exact reality (avoiding the principles that define a model [238]) it might not be enough in order to obtain a full range of knowledge for the system. The system should exploit this model in order to obtain this full range of knowledge about this reality. That is, the one that concerns the experience.

From this point of view, there are essential differences between models and the experience through the exploitation of these models. This might be better explained from the perspective of computer science: I understand that it is not a model of an object what provides essential knowledge to the system in order to obtain experience. It provides just a reference for this experience.[5] This only allows for the instantiation of this object, and for using it. What is essential for experience are those relationships created on the basis of the interaction between the instance of the object and the environment of the system where the instance works.

It is clear that a model can be exploited without the need of being experienced. What seems to be relevant, is that it has to be experienced by the agent to achieve interpretations (i.e. semantic) about this model.

---

[5]The large study beyond the state of the art makes me assume several other references for experience.

2.7 Self-Adaptive Systems

Our main focus deals exclusively with the study of the influence that emotion causes inside of a system. We analyze the key foundations of the emotional work that clearly improve the adaptiveness in life systems, and study models in order to obtain a partial similarity in artificial systems.

We should firstly distinguish the idea of *adaptation when it refers to evolution* from *adaptation when this is linked to the environment*. The former concerns the process of evolution between generations which allows for new individuals with improved resources adapt better to the common generalities of their environments. The later refers to the management of those resources, in order to survive when these environments become dangerous. The state of the art argues that emotion appears within the first scenario, and that it is used within the second scenario.

Biological agents use feedback strategies in order to adapt themselves to their environment. This environment may be understood either as a situational or a physical environment. They are autonomous systems with such a complexity, that they require decentralized organizational and operational structures to allow efficiency in their operativeness. The emotion seems to provide a set of functional and organizational patterns, that become essential grounds in order to manage the inner complexity of biological agents.

Under the vision of [9], feedback loops are the tool by which the system senses the environment, executes control strategies and execute actions to ensure stable dynamics—i.e. bounded disturbances have to induce bounded errors.

Under Brun et al. [37], self-adaptive systems have in common some essential features: (a) design decisions that have to be moved towards runtime control dynamic behavior, and (b) the work of an individual system that reasons about the state of the system and the environment [37]. He as well as numerous other authors (see the roadmap in self-adaptive systems research by Di Marzo Serugendo et al. [63]), argue feedback loops strategies in order to provide a generic mechanism for self-adaptation. Commonly, it is referred as the *autonomic control loop* and involves the following four activities: (a) collect, (b) analyze, (c) decide and (act). The operation is started by sensors that collect data from the executing context and the context. These data is arranged and stored for future needs (model of past and current states). After a diagnosis process in order to analyze the data, the planning attempts to predict the future, and decide the following actions on the environment through effectors.

Di Marzo Serugendo et al. [63] hold that self-adaptive systems must be based on feedback strategies. This feedback behavior of adaptive systems is realized by means of its control loops, and it is argued to be an essential feature of these systems [63].

As it has been previously introduced, software engineering uses models as a well-established form of information integration. The concept of adaptation is closely related to emotion, since emotion is considered an essential precursor and trigger for adaptiveness. As it happens in life systems, the idea of adaptive systems is used in engineering in relation to how systems might optimally deal with their environments. That is, how to maintain their resilience by improving their feedback mechanisms.

This becomes the ability of self-managing at runtime, to manage a situational state caused by some change. From the perspective of Vogel and Giese [264] this is a question of how the system configures and reconfigures itself, continually tuning and optimizing, protecting and recovering while keeping its complexity hidden from the user [4, 264].

As we have previously said, this thesis will assume the task of modeling cognitive processes by holding the principle of *Model-Based cognition* argued by Sanz et al. [230]. Thus, we will focus on those runtime self-awareness paradigms of relevance for our approach such as self-adaptiveness, model-centric systems and cognitive architectures.

There have been formalized metacontrol solutions for *Self-aware Autonomous Systems* Hernandez Corbato [106] within the framework of the *ASys Project* that surrounds this thesis. Part of this formalization is built under the named *Epistemic Control Loop pattern (ECL)*, a model-based cognitive control loop. The based structure of an ECL is organized by means of two artifacts: a feedback control strategy, and a model-based predictive control (i.e. model-based control) (see Hernandez Corbato [106]). And assumes the use of *models@runtime* in order to reason about the system and its operating environment at run time (see Daubert et al. [58]).

The artifacts used as the knowledge within ECL [106] are *runtime models*, a causally connected "self-representation of the associated system that emphasizes the structure, behavior, or goals of the system from a problem space perspective under the view of Blair et al. [24].

Within this environment, Vogel and Giese [264] proposes the separation between the adaptation engine and the adaptable software. Self-adaptation is conformed by (a) the adaptable software which becomes fully aware of the domain logic, and (b) the adaptation engine that implements the adaptation logic as a feedback loop. The work of Vogel and Giese [264] approaches *EUREMA*, a domain-specific modeling language and a runtime interpreter for adaptation engines [264]. Commonly, engines are feedback loops, and describe the adaptation by (a) runtime models that represent the adaptable software, and (b) activities of analysis and planning that use these models. Vogel and Giese [264] proposes the *megamodel* artifact in order to systematically address the interplay between runtime models and adaptation activities. Since runtime models are the knowledge within the feedback loops, these *megamodels* are kept alive at runtime in order to be interpreted, dynamically adjusted to adapt feedback loops and directly executed to run such engines.

The adaptation engine used by Vogel and Giese [263] assumes an external approach (i.e. the architectural approach) based on the work of Salehie and Tahvildari [225] (Fig. 2.6a), that splits the *self-adaptive software* into the *adaptation engine* and the *adaptable software* controlled by the former one [263]. The adaptable software realizes the *domain logic*, while the adaptation engine puts into effect the *adaptation logic* as a feedback loop. A more detailed view of this external approach is provided by using the *MAPE–K cycle* of [123] (Fig. 2.6b). This is a refined adaptation engine with a structure of four activities sharing a common knowledge base: Monitor, Analyze, Plan and Execute–Knowledge.

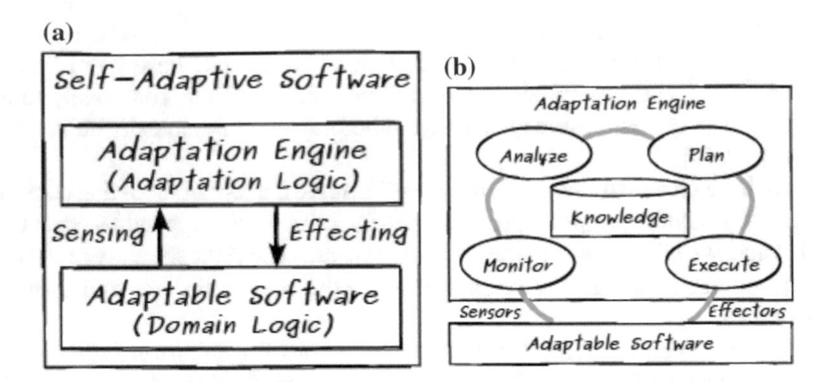

Fig. 2.6 The vision of self-adaptive software under Vogel and Giese [263]. (*Source of figure* Vogel and Giese [263])

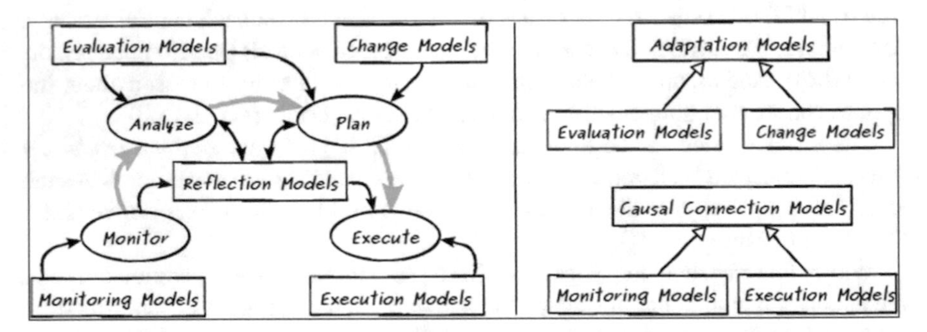

Fig. 2.7 Runtime models perspective by Vogel and Giese [263]. (*Source of figure* Vogel and Giese [263])

Vogel and Giese [263] approaches a broader perspective and propose a new categorization of runtime models, under an extended version that considers all models that are used on–line by any adaptation activity of a feedback loop (see Fig. 2.7). This way, *Reflection Models* reflect the adaptable software and its environment.

The growing of abstraction in this work is assumed under the employing of coordinated multiple feedback loops that are, independently and concurrently working, while at the same time are in continuous interaction. This coordination is assumed under the work of layered architectures that organizes feedback loops in layers (see Fig. 2.8). Feedback loops at higher levels will control the adaptation engines (i.e. feedback loops) at the layers directly bellow (the lowest layer directly controls the adaptable software). And *EUREMA* is approached in order to model these feedback loops and the required reflection models.

Since *AGSys* is assumed as a Model-based cognitive agent, the agent program can be thought of as the wrapper of multiple modules of adaptive software. And their implementation can be interpreted as self-adaptive modules by using this Vogel and Giese [263] approach.

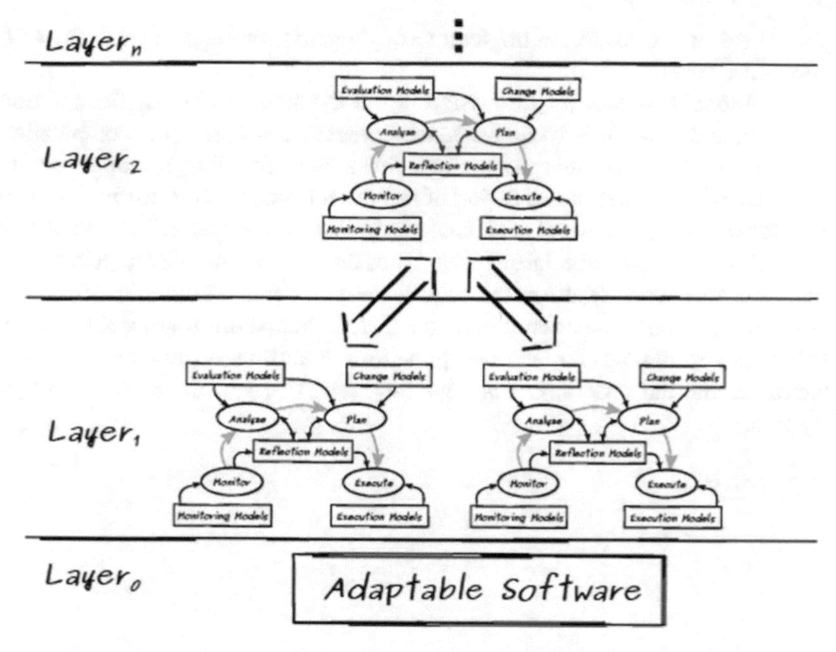

Fig. 2.8 Layered architecture for an adaptation engine by Vogel and Giese [263]. (*Source of figure* Vogel and Giese [263])

Consequently, this thesis will refer the use of these adaptive engines proposed by Vogel and Giese [263] when model self-adaptive attributes in *AGSys*.

2.8 Conclusions

This chapter has introduced the use of models as they have been used along this thesis. The wide problem which constitutes artificial emotion requires a large number of models of quite different nature. The three major problems to model emotion are (a) the design of those processes that are responsible of assessing and that afterwards becomes the background for experience, (b) those processes that—once obtained the ground for experience, provide the system with the observation of feeling, and (c) how to adjust the self-adaptive capabilities of the artificial system to the new conditions that impose the observed situational state.

Regarding the the design of processes, the requirement of a finite *carrier process* is conceived in order to provide a continuous emotional assessment. And the reasons regarding finiteness have been provided in this chapter.

Additionally, it appears as if the emotion based objectives (those that drive the emotional behavior) are closely related to the evolution of the emotional process. So, emotional drives are related to the feature of *finiteness* through the stability of the emotion–like goals.

An additional problem for discussion is the dynamic feature of the emotional phenomena, and how these dynamics can be integrated as part of the computational process. Since dynamics emerge from unstable states regarding emotional goals in life systems, this can be assumed as part of these emotional goals in artificial systems.

The issues of perception and models of experience and self are critical problems in the general challenge of artificial emotion. Since this is a long term attempt in artificial intelligence, this work assumes the requirement of relational entities in order to allow the integration of experience in models. We hold that this relational knowledge will provide sustainments for later interpretations regarding the relation between the environment and the system, that is, regarding the effects of the environment over the system.

Chapter 3
State of the Art

This thesis has needed a particular purpose of understanding emotion science from the wide range of perspectives that we can find in the state of the art. Any research question requires to study the particular condition of methods and sources regarding the topic at the specific time of the investigation. The state of the art is an essential tool to define the origin that might solve the question, and support the studied solution later. Regarding this thesis, this study has been the cornerstone of our research. Not only were we needing knowledge about the state of the investigation in emotion science, but also the very requirement of understanding the essentials of the real phenomenon on the basis of the multiple alternatives available about this topic.

3.1 Introduction

State of the art in emotion science is not a question of parallel results over time; the problem of emotion science is that it branches along every question that the phenomenon establishes. And it establishes a new wide branch with its own state of the art. The essential question in our research has been the search for understanding each element of the problem, in order to match analogies in artificial systems.

Most conflicts arise from a simple correlation between the biological emotion and the functional segregation of artificial systems in emotion deployment. Computational solutions integrate isolated parts of the emotional process, under the constraint that imposes how the theories of reference are built. And I have constructed the required base of knowledge by putting theories pertaining to different branches, with the aim of incorporating each branch of knowledge under a permanent structure that might help to solve a punctual question.

M.G. Sánchez-Escribano, *Engineering Computational Emotion—A Reference Model for Emotion in Artificial Systems*, Cognitive Systems Monographs 33,
DOI 10.1007/978-3-319-59430-9_3

So that, this chapter will respond to the clarification of the large path followed concerning knowledge searching and under the logic of a general investigation. It is a summary of a part of the whole work made regarding the general study of this thesis. When studying this wide range of emotion science references, the objective has been the construction of a set of essential antecedents—either in emotion science as well as in artificial emotion applications. However, I have mainly focused on building a comprehension of the phenomena, as well as the problem and solutions described in artificial agents. To this end, I have searched within a wide range of works without discarding any perspective showed within them.

I have however assumed that baselines and principles of emotion, come from relevant sources that study the real phenomena. And from this perspective, I have built a theory by joining several elements of different theories. Consequently, the next chapter will describe the range of theories that have supported our vision regarding this thesis. So, I will not describe herein more than an essential portrait, since the following chapter will describe the foundations and baselines for the present work.

This chapter will thus respond to the logic that preceeds our research, letting the following chapter describe the logic that supports our solution. Nevertheless, the following chapter must not be considered as state of the art, but as part of these argued conclusions within this thesis—since these conclusions are the baselines and principles that I have defined in order to build our theory.

3.2 General Overview

The structural and functional studies from the perspective of biological emotion can be summarized into two major approaches: (a) the *physiological perspective* that focuses on the structural basis within the brain, and (b) the *psychological perspective* that addresses the emotional phenomenon within the agent. This section will describe these two perspectives from a historical point of view over time, with the goal of showing those main conclusions that afterwards are commonly accepted and the principles from which they were defined.

The analysis of physiological bodily changes during an emotional process, provides sets of data to explain the physical origin of emotions as well as the extent of emotion. The classic vision of James [118]—detailed later by Lange and James [133], shows emotion as a mental representation of body responses to relevant stimuli. The emotional process was also refined by cannon who argues a simultaneous approach to the physiological change. He argued that the subsequent effects of the physiological change have their origin in the hypothalamus region of the brain [43]. Analogous to his conclusions, Papez [201] in his initial approach suggested the distributed nature of the emotional processes in his dual circuit model. Accordingly, the works of Klüver and Bucy [128], MacLean [155] and LeDoux [139, 143] showed that this structure is not so clearly delimited in the brain as it might have seemed originally.

Subsequent studies ratified this with techniques in neuroimaging. These studies demonstrated that the effects appearing while fear is felt, are those exact effects that appear during the unconscious perception of this basic emotion (see Morris et al. [177]). Additional works in this line, suggested the involvement of several other [deeper] subcortical structures in emotional processing. Evidences such as the one of *the case of Phineas Gage* [55] showed the relevance of the cortex area in emotion processes, as well as several other upper structures. Also, empirical studies made on the basis of brain lesions—concerning the role of the *orbitofrontal* and the *ventromedial* cortex, showed that these structures are playing an essential function in (a) the evaluation of the external information regarding new events (i.e. new outward stimuli), as well as (b) the internal information regarding bodily sensations and emotional states (i.e. new inward stimuli). The studies made by Davidson et al. [60] and Bush et al. [38] confirmed these hypotheses, and showed that the *anterior cingulate cortex* is related to the expression of emotion and to the conscious experience.

From these results a distributed operation of the emotional system rather than a centralized work was proposed. It seems that the emotional phenomenon is not located in a particular brain structure—from which a centralized control is performed. Rather, those processes that are sustaining emotion are complex functional chunks in a simultaneous, federated and distributed work. This complex operation as well as the nature and form of emotional processes, are major challenges in order to model the emotional phenomenon.

One major issue is how this set of interconnected processes—that occur at several levels and different structures within the brain, are leading to abstract complex processes such as those of cognition, emotion and consciousness.

Clinical studies conducted by Damasio et al. [55] and some others using neuroimaging such as in Davidson et al. [60], confirmed the modulatory role that the *prefontal cortex* plays over several other brain structures. This is related to the adaptive nature of the emotional response in Miller and Cohen [169] and Ochsner et al. [194], one of the multiple facets related with the emotional operation. Prinz classified the common perspectives of emotion study: (a) the feeling theories related to the conscious experience of emotion, (b) the somatic theories regarding the physiology of emotion, (c) the behavioral theories linked to the action tendencies concerning the emotion, (d) the cognitive theories connected to the modulation processes concerning the emotion, and (e) the pure cognitive theories related to the thoughts and the emotions [212]. This work described the relevant idea of a multiple role of emotion in life systems.

Nowadays the idea of emotion as a cornerstone for intelligence is widely accepted (Goleman [91]). This idea was introduced in the early eighties as a feature of systems that would allow them to intelligently use their resources. See Sloman and Croucher [241] who argues that "any intelligent system with multiple motives and limited powers will have emotions". According to Minsky, this is not a question of whether emotion should emerge in machines, but a question that considers if machines could be intelligent without emotion ([170]), a shared idea by Cañamero [41], who argues that emotions are a key base for intelligent behavior.

The psychologist's point of view refers to the study of emotion as an intrapersonal process Lewis [149]. The psychological perspective provides some generalities about the emotion functionality such as the modulatory function that affects the agent's actions concerning its adaptation to the context. This perspective commonly addresses the analysis of those processes that allow for optimized responses, focusing on adaptation and self defensive behaviors in life systems. And they commonly address the analysis of the role that emotions play concerning the influence on cognitive processes such as perception, learning, decision making and communication among other.

One of the aspects that draw more attention in the earlier stages of this research, was the question of how emotion is caused (see the works of Strongman [249] or James [118]). A second issue was the determination of the mechanisms required to give rise to this phenomenon (see *The Four systems for emotion activation* of Izard [116]). Works related to this latter question have defended the viewpoint of thinking about emotion as a state, resulting in relevant works of characterization, states differentiation and features definition among other (see Ekman [66, 68], Darwin [57], Oatley and Johnson-Laird [192], Emde et al. [72], Panksepp [200], Arnold [5] or Plutchik [208]).

Earlier theories come from philosophy, from doctrines of affections established by Plato (427–347 B.C.) and theories about essential emotions of pleasure and pain by Aristotle (384–322 B.C.) who thought of emotions in a way opposite to that of Plato. Affections have been studied along the history: from *Patristic* and *medieval doctrines*, to the *affective psychology* during the Renaissance. It is in s. seventeenth century that emotions are conceptualized as a *system of passions*. Examples include the work of Malebranche-who identified seven moments that together make up the structure of the passions; Hobbes-who identified emotion with material locomotion; and Spinoza-who showed less interest in the bodily causes or consequences of emotion, and focused on the explanation of mind operation. Lastly, it was Rene Descartes, in his conceptualization theory, argued emotion arised from the close and intimate union of body and mind (see Gardiner et al. [84] and Zalta [275] for detailed descriptions of these theories).

In artificial systems, a major area of interest is the study of emotional expression. Some papers have focused on basic emotions (see the studies about basic emotions of Izard [116], Camras [40], Oatley [189] or Ekman [67], whereas other relevant work has analyzed topics such as culture and emotion communication (Izard [115]). However, nowadays, the *appraisal theories of emotion* have become relevant and stay closely connected to the matter of emotion science because of the interest in understanding the emotional assessment. The term *appraisal* is attributed to Arnold [5] who understood appraisal as a process of evaluation related to assessments of relevant situations. This is a broadly accepted conceptualization, even though there is no agreement regarding the origin of this process [80, 237].

There are two pertinent questions within the *cognitive approach of emotion*: (1) the *stimulus relevance*, and (2) the arising of emotion on the basis of individual's concerns (see Oatley [189], Ortony et al. [199] and Johnson-Laird and Oatley [120] among other). The *cognitive approach of emotion* has also inspired works addressed to build computational theories of emotion. Some of these computational theories have been of great relevance in the state of the art of artificial systems (see for example the essential works of Ortony et al. [199] and Bartneck [16]).

Finally, there are many questions that are hard to solve and which still are debated such as that of the *embodiment of emotion*. Whilst authors such as Herrera et al., James and Damasio agree the viewpoint of *embodied emotion*, there are relevant authors such as Fox who holds that "there is a little doubt that emotions are embodied phenomena".

The implications of emotion in cognitive and subjective issues of the mind covers a large number or works and a wide scope of subjects. Scherer [235] argues that emotions are important tools for mediating interchanges between organisms and environments, and serve to decouple stimulus and responses throughout a threefold process: (1) evaluate situations (i.e. organism needs), (2) prepare psychological and physiological responses, and (3) aid in communication and behavior. Bower [29] argues the implication of emotion in memory and learning, a point of view shared with LeDoux [141], Gardner et al. [85], and Goleman [90]. This latter author referred to this implication of emotion as a cornerstone of intelligence, whereas Damasio et al. [55] provided and equivalent statement with respect to reasoning.

The *subjective experience of emotion* is an old issue that also has been widely studied in the state of the art. Giving an example, even when Ekman [66] described his classification around six essential emotions, he also assumed this perspective. He suggested that, even when it is not the only effect, the experience of emotion includes sensations that are a result of some feedback concerning changes occurring within the response system of the agents [66]. The *subjective experience* includes-but may not be limited to-*sensations* that are the result of some type of feedback that comes from the source of changes that occur during the emotional phenomena. Damasio [54] provides a relevant model in order to describe the experience of *subjectivity* and *feeling*[1] [54]. Damasio [54] however argues the requirement of *self* in order to *feel that you feel*, an idea that is in accordance with Ettinger [73] that argues a *self circuit* in the brain as the seat of *feeling*. Fox [77] argues that "feelings are the conscious mental representations of emotions and moods".

Emotion science provides a wide range of essential questions that regard emotion, emotion schemes, moods, feelings, conscious and unconscious feelings, temperaments or affect among several others [77]. There have been attempts to describe exactly the nature, scope or meaning of emotion such as those of Wierzbicka [269], and Kleinginna and Kleinginna [126], and there are lot of authors that argue for the search of proper understanding of vocabularies such as Harre [102].

[1]This theory of Damasio [54] will be detailed in the following chapter since it is a cornerstone in our thesis.

3.3 Trends in Biological Emotion Research

There can be no doubt that the twentieth is remarkable in emotion research, which keeps on improving to this day. In fact, there are so many new theories defined and so many advances in emotion science, that it is difficult to enumerate the complete list of models and works. However, I think I have managed to whittle it down to those that have had the greatest influence on emotion science.

Two large lines of research during the early ages of emotion science can be distinguished: (a) the physiological perspective, which states the structural basis of emotion within the brain, and (b) the psychological perspective, which explains the characterization of emotion within the system.

3.3.1 Research on the Neural Bases of Emotion

The analysis of emotional bodily processes conforms the knowledge base on top of which several theories are built, most of them focusing on the explanation of the origins of emotion. As has been previously introduced, James [118] understood emotion as a mental representation of bodily changes under relevant stimuli [118]. Subsequent works Lange and James [133] incorporated new additional components to these physiological changes, but they do not explain how the environment induces these changes or how different emotions might be featured from similar physiological changes [133]. Early in 1927, Cannon [43] refuted the theory of James [118] by arguing a non-direct causality in emotion cause-effect [43]. However, the initial theory was rapidly modified by Bard [14], who argued a simultaneous occurrence of emotion and bodily changes after an event, considering the hypothalamus as the central structure for the emotional brain [14]. Later in 1937 Papez [201] agreed with the common work of Cannon–Bard, and considered this work as a step towards the knowledge of the hypothalamus structure [201]. Papez [201] denied a centralized operation of emotion and proposed the limbic system as part of the emotional control system [201].

Even if not completely accepted nowadays, the *Papez's cirquit* was an essential contribution for the explanation of the neural paths of emotion. The main contribution of Papez [201] was the differentiation between the *thought channel* and the *sentiment channel*. The former was argued to be the responsible of endowing sensorial information through the thalamus towards the lateral neocortex to become perception, thought and memories. The sentiment channel was argued to be responsible of the subjective experience through thalamus to hypothalamus and cingulate cortex. Some time later Kluver and Bucy [128] studied those brain areas that were the primary cause of visual hallucinations, and this work results in an essential cornerstone for the work of McLean [164].

McLean [164] recovered the discussion regarding the theory of Papez [201] and proposed the *visceral brain* as the base of the emotional experience. The work of this *visceral brain* was conceived as a federated operation between (a) the limbic system that produces a visceral response, and (b) the hypothalamus that structures the information [164]. While McLean [164] argued this integration within the hypothalamus, Papez [201] suggested however the cingulate cortex, and proposed in later theories the integration of amygdala, prefrontal cortex and the septum as centers for survival and self-protection. It should be pointed out now that modern studies (such as those conducted by LeDoux [139]) have proved that the limbic system is not a bounded structure.

LeDoux [142] is the paradigm regarding the studies on amygdala and how this structure influences *fear conditioning* and *defensive responses*. He established a set of conditioned and non conditioned stimuli in order to assess fear conditioning [142]. Once neuroimage was available, the concurrent and proportional activation of the amygdala as a result of fear perception could be proved. Morris et al. [177] arrived at this same conclusion in the course of an unconscious perception of the stimulus, and proposed the activation of two additional kernels (i.e. superior colliculus and pulvinar).

All these studies led to an essential conclusion: there exist subcortical paths in order to detect emotional phenomena. Harlow, the doctor of Phineas Gage, became aware of the personality changes showed by Gage after a prefrontal cortex lesion. Damasio et al. [55] extended a posterior work beyond Gage's case, and argued the influence of the prefrontal cortex on the brain structures of emotion, an idea that has been afterwards proved by neuroimage based studies by Davidson et al. [60]. It seems as if the prefrontal cortex might modulate (rather than mediate on) the emotional response, something related to the adaptive feature of emotion (see Miller [169] and Ochsner et al. [194]).

The Phineas Gage case led the conclusion that orbitofrontal and ventromedial cortex were responsible of these personality changes. The ventromedial cortex is argued to be a secondary association area that integrates different types of sensor and somatosensorial cues in order to maintain systemic equilibrium. Thanks to this secondary association, the external information (regarding new events) as well as the inner information (regarding bodily sensations) are assessed. Papez [201] had already studied the association of the anterior cingulate cortex with the conscious experience, but the work of the conscious experience, but Davidson [59] and Damasio et al. [55] confirmed the hypothesis (arguing additionally that the anterior cingulate cortex might be related to expression and conscious experience of emotion). Both works (i.e. Davidson [59] and Damasio et al. [55]) establish a differentiation between the *affective* and *cognitive* parts of this region. The *affective* part being related to the autonomic nervous system by taking part in control activation during the emotional experience, and the *cognitive* part being related with cognitive processes by taking part in the selection of the emotional response.

As a result of all these theories, it seems as if the emotional phenomena did not come from the source of a central structure from which a centralized control is made. It rather seems as if processes related to emotion were simultaneously triggered, being distributed among a set of small functional modules that work in a federate way in order to maintain a systemic equilibrium. However, the major challenge regarding modeling is not the operational complexity of these brain structures, but how these complex processes are interconnected in order to result in cognition, emotion and consciousness.

3.3.2 Psychology and Emotion Research

Psychology constitutes an essential and wide field in the search of principles that explain cognition, emotion and consciousness. Within this field, the state of the art describes emotion on the basis of observed effects. By and large, there is a major agreement about the influence of emotion on adaptiveness and self protection. Nevertheless, they are not the only accepted features. The role of emotion in cognitive processes such as learning, perception, assessment or decision making, the associated bodily changes, the close collaboration with communicational processes or the subjective experience of emotion, are some main issues in this field.

Prinz [212] classifies the wide field of emotion theories in five categories-depending on the emotional episode that constitutes the basis of the study: (a) *feeling theories* regarding the conscious experience, (b) *somatic theories* regarding bodily changes, (c) *behavioral theories* regarding action tendencies, (d) *processing mode theories* regarding cognitive processes modulation, and (e) *pure cognitive theories* regarding thoughts.

Under some general acceptation, psychology establishes a general division between *basic emotion* (which induces visible changes within the system) and *complex emotion* (which affects cognitive processes and emotional experience). They are also differentiated states, those that are more exclamative (intense and punctual states), from those related to the idiosyncrasy of the system (moods and affective phenomena).

1. *Basic Emotions*
   Ekman [66] proposed one of the most relevant theories of emotion categorization on the basis of emotional expression, and improved this theory later in a shared work with Friesen (see Ekman [68]). There are however a wide amount of theories on emotional conceptualization and discretization (see Darwin [57], Oatley and Johnson-Laird [192], Emde et al. [72], Panksepp [200], Arnold [5] or Plutchik [208]). It leads to a common pattern of basic emotions that however, is featured by the absence of bounds concerning meaning.
2. *Complex Emotions*
   They state a more complex question with fuzzy conceptual delimitations. The experimental base for their study does not offer objective results to be measured with some type of metric. This absence of data is not only realized in order to assess the emotion influence on a punctual cognitive process. It is also reflected in discussions in order to clarify terminology such as *mood*, *temperament* or *trait* among several others.

Watson and Clark [268] analyzed the terms *emotion*, *mood*, *temperament* and *trait*, arguing differences not only among related states, but also in how they affect the final response of the system within similar environments. Isen and Simmonds [114] studies focused on cognitive flexibility related to positive or negative states of *mood*. By and large, these studies show that positive or negatives values concerning *mood* may have a significant influence on—or even determine—different responses under the similar circumstances. There are a lot of empirical works that reveal different tendencies in cooperative tasks, as the one by Hornstein et al. [110] to study this behavior during a play. Smith and Lazarus [243] show some doubts regarding the causality of *moods*, and justify motivational and cognitive conditions as the responsible of supporting differences between intense emotions and moods.

Currently there is a general consensus in accepting a mutual influence—even integration—between emotional and cognitive processes (assuming that each author presents arguments from quite separate perspectives). Hud [1] argues that this relationship is more visible in processes of decision making, and justifies her theory by modeling the effects of emotion on cognition (by using a cognitive-affective symbolic architecture). This influence has been studied on several other cognitive processes such as perception (see Cytowic [52]), on memory (see Bower et al. [28]) or on motivation (see Izard [116]) among others.

One major area in emotion science is the *Cognitive Appraisal theories of Emotion*. The term *appraisal* was introduced by Arnold [5] by means of which she described an emotional assessment linked to a judgement, which concerns some relevant situation. However, there are still no conclusions about the source from which this evaluation emerges [237]. What is more widely supported is the idea of *appraisal as a conscious evaluation*, as a means to obtain meaning (see Lazarus [137]). *Appraisal theories* postulate a continuous assessment about a situation, and emotions are represented as vectors in a way to represent components of emotion and values of these components throughout different emotional situations. Scherer et al. [237] divide valuable components of appraisal into four wide categories: *implication, relevance, norms and coping*. A work of major influence is the one of Ortony et al. [199] that is based on the the search for "what are the distinct emotion types and how are they related to one another" in order to obtain a cognitive theory. This theory defined a taxonomy named the *Global Structure of Emotion Types* by assuming that "there are three major aspects of the world, or changes in the world, upon which one can focus, namely, events, agents and objects". This way, the hard problem of perception and subjective experience is moved away from the problem of emotion.

The explanation of the *subjective experience of emotion* is featured by multiple works focused on the task of explaining different—or even isolated—stages of the phenomenon. Likely, the theory of Damasio [54] is among the ones with major acceptation in the state of the art regarding this question. This work states an emotional control in a transversal organization with the objective of giving an explanation to the *experience of feeling*, arguing influences on reasoning and decision making as a consequence of this experience.

As a general issue in emotion science, existing theories place emotion at a crossway of roles in order to justify adaptation [79]. Even when there are common points of agreement, the question of building an universal theory is extremely complex. It is closely related to the subjectivity of data, to the empirical works that support each theory, and to the absence of uniformity regarding perspectives, disciplines, or terminology under which every question has been studied. This particular condition is constituting a large barrier, even with the objective of understanding how works ensemble one to each other.

3.4 Trends in Artificial Emotion Research

The state of the art in artificial emotion is featured by heterogeneity regarding motivations, frameworks, objectives and methodology. Even with similar foundations in order to support each work, there is a heterogeneous collection of results for the same objective or question. By and large, works are focusing on isolated parts of the emotional phenomena, without additional considerations regarding the interaction—or even overlapping—circumstances under which real processes work, and that are hardly influencing the real phenomenon.

Zhang et al. [276] present an analysis of the state of the art around computational emotion by establishing six critical aspects by which they argue emotion should be represented: (a) even when models are associated to a set of emotion types, they should be adaptable in order to explain additional emotions, (b) models should allow for simultaneous emotions, (c) models should allow for different emotions for two systems within the same context, (d) system needs the ability of mapping emotional situations into rules, (e) models should allow for "other systems emotional representation", or even more, of representing emotions of the group, and (f) emotional responses should emerge from the source of emotion and system goals.

The works of Ekman [66] and Ortony et al. [199] provide two relevant frameworks within which a lot of foundations for computational emotion are built. The center of interest is the creation of environments of knowledge aiming at analyzing questions such as character, personality, empathy based communication, emotion elicitation or humoral changes (among others) in artificial systems (see Reilly [216], Elliot [69], Koda and Maes [129], Ortony [198], Plutchik [208]).

There are a set of general features commonly argued as essentials in emotional models such as (a) cognitive knowledge on the basis of theories of value and feeling, (b) the integration of patterns for personality and action tendencies in order to improve functional modulation, (c) the integration, detection and understanding of emotion to allow for meaning, (d) the integration of new meaning through the arousal and appraisal processes in order to modulate behaviors, and (e) the challenge of goals in order to implement processes of attention. However, the state of the art shows a broad range of approaches aimed to cover specific stages of the emotional phenomena, usually focusing on intelligence improvement, emotional expression, motivation and decision making.

This following summary is arranged according to the objectives of this thesis. The main purpose of this survey is to describe the state of the art within the context of emotion science in artificial systems. A context that aims at searching for computational supports in order to obtain intelligent capabilities in systems.

3.4.1 Emotion in Artificial Intelligence and Robotics

Ortony [198] assures that an appropriate response of an artificial system that implements emotion "depend[s] on implementing a robust model of emotion elicitation and emotion-to-response relations". Picard [205] argues an essential role of emotion in cognitive functions and intelligence and, she even argues a critical role of emotion on decision making, human interaction and perception [205]. From this perspective, Picard and Picard [207] put forward a strong argument to support the requirement of affective computers if intelligent machines are required, and even gives reasons to support the idea that "a machine wouldn't pass the Turing test unless it is also capable of perceiving and expressing emotions" [207]. And Loia and Sessa [150] argue that emotional research in non-biological autonomous agents "is not new, and it even references some ideas about the requirement of emotion for truly intelligence" [150]. Canamero [42] even holds that "modeling emotions in autonomous robots can help towards understanding human emotions as sited in the brain, and as used in our interactions with the environment and emotions in general".

By and large, modern approaches relate emotion and intelligence as a whole functionality with subtle interactions one with each other. However, instead of providing improvements regarding the intelligence of the systems, works in this area widely follow the challenge of a more natural human–machine and human–robot interaction (see Leite et al. [145] for a survey in social robotics, and Cabibihan et al. [39] for conclusions in human–robot interaction), i.e. works aimed to approach skills related to communication, empathy interaction and collaborative tasks.

From the perspective of communication capabilities, the empirical work of Reeves and Nass [215] showed some relevant features of human's emotional responses while interacting with machines [215]. Humans showed more ability to understand and share communication with machines if machines were showing emotion. Emotion research in *human computer interaction (HCI)* and *human machine interaction (HMI)* are challenging objectives that are currently resulting in successful solutions. The area of *affective computing* shows a wide amount of works with relevant results in artificial systems. The main actions in *affective computing* are driven towards the recognition and representation of emotional patterns (see Picard [205], Yu et al. [274] and Kim et al. [125]). Each day more, we can find entertainment equipments, plays, virtual agents, intelligent interfaces or robots which are able to cause emotions in humans (see for example Tamagotchi ®,[2] The Sims ®,[3] Nareyek [179], Bell et al. [20], LEGO [144]).

[2]http://www.bandai.com.

[3]http://www.thesims.ea.com/us/.

The *Robotics* field—and even more intensified the field of *Social Robotics*, is continuously improving designs in order to obtain prototypes of robots with similar behaviors to those showed by humans. Bye and large, these prototypes aim at different objectives within the fields of empathy and easy communication (see Breazeal et al. [33]). The *Robot Kismet* is probably one of the most widely known prototypes, closely related to the work of Breazeal et al. [32] about emotion and motivation (see Breazeal et al. [32, 33]). *Kismet* is a prototype of autonomous robot developed to model a motivational system to regulate human–robot interaction (see Breazeal [31]). *Kismet* was designed to have social interaction with humans, and provided with a broad variety of facial expressions to display or mimic emotions (see Fig. 3.1).

MIT COG was a project at the *Humanoid Robotics Group* of the *Massachusetts Institute of Technology (MIT)* developed with the goal of understanding how humans work (see Brooks et al. [36]). The early hypothesis was that human intelligence requires experience from the source of interaction. The main feature of *COG* was the ability to appropriately understand and share the humans social environment. This way, through the interaction with humans over time, *COG* would be able to learn socially under observation of emotion, wishes and objectives (Fig. 3.2b).

Waseda University's Sugano Lab created a sequence of emotional communication robots as a subsequent work of *Hadaly-2* and *WENDY* previous projects. *Robot WAMOEBA* is an autonomous robot which has an emotion model aimed to maintain a friendly communication with humans (c.f Ogata and Sugano [196] and Ogata et al. [195]). The emotion model draws the endocrine system of humans by using four hormone-like parameters to adjust inner conditions of the robot such as motor output, sensor gain and temperature. This communication aimed at the empirical study of emotion influence in cooperative tasks (see Fig. 3.3).

Another line of research is that aiming to establish facts—and approach conclusions—regarding how personality factors affect system behavior. Among the

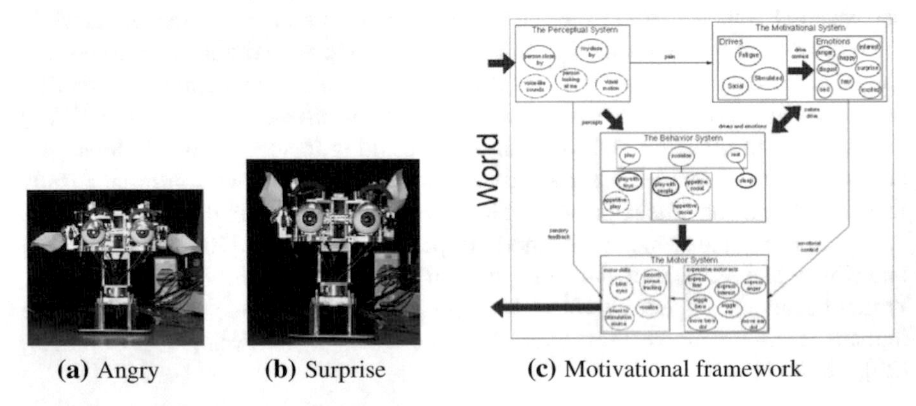

(a) Angry **(b)** Surprise **(c)** Motivational framework

Fig. 3.1 Kismet is an autonomous robot developed by Breazeal [32] at MIT Artificial Intelligence Laboratory: **a** Kismet displays an angry expression, **b** Kismet displays surprise, **c** Framework for building the motivational system. *Source of figures* Breazeal [32]

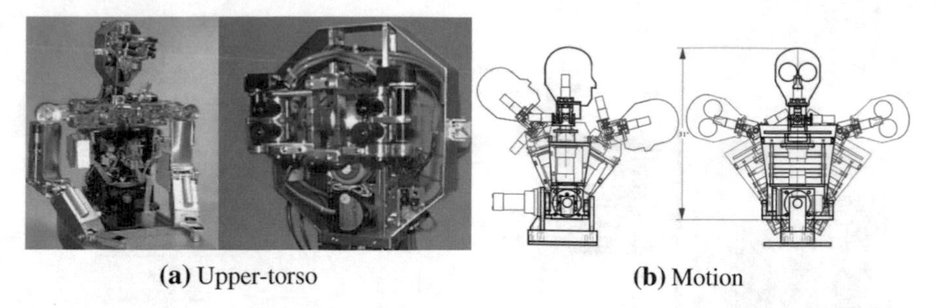

(a) Upper-torso **(b)** Motion

Fig. 3.2 Cog is an upper-torso humanoid robot: **a** cog is a platform with twenty-one degrees of freedom and sensors enough to approach human movements and senses, **b** range of motion for the neck and torso. *Source of figures* Brooks et al. [36]

Fig. 3.3 The sequence of "Wamoeba" robots from *left* to *right*: Wamoeba-2, Wamoeba-2R, Wamoeba-2Ri and Wamoeba-3

required capabilities in artificial models, there is an increasing interest for integrating *understanding-like* abilities such as those showed by humans (see Hawes [103] and Ghasem-Aghaee and Oren [88]). And commonly, the used methodology is the integration of patterns that draw persistent features of personality. These patterns are intended to interact with emotional models, in order to assess the influence on the behavior of systems.

A well known example is the platform developed by the *Robotics Institute of Civil Aviation University of China* named *Fuwa* (see Fig. 3.4), a customers attention platform able to elicit emotion. Under the assertion that social interaction and communication are hardly influenced by emotion, the objective was a robotic system capable of deploying emotional communication in order to change final behavior if it were required (see behavior if it were required (see Qingji et al. [213]). *Fuwa* is a service robot aimed to provide entertainment, which captures the environment through a CCD camera, ultrasonic and microphone sensors in order to obtain face, obstacles and voice detection. The emotional mechanism is based on *PAD Emotional State Model* Mehrabian [165] and integrated together with a model of personality based on *The Five Factor Model* Digman [64].[4] The final experiment aimed at the

[4] *The Five Factor Model* Digman [64] is a well known framework in psychology which is widely used in robotics. It is an essential supporting structure that regards personality influence, to understand the capacity of personality's factors to have effects on behavior.

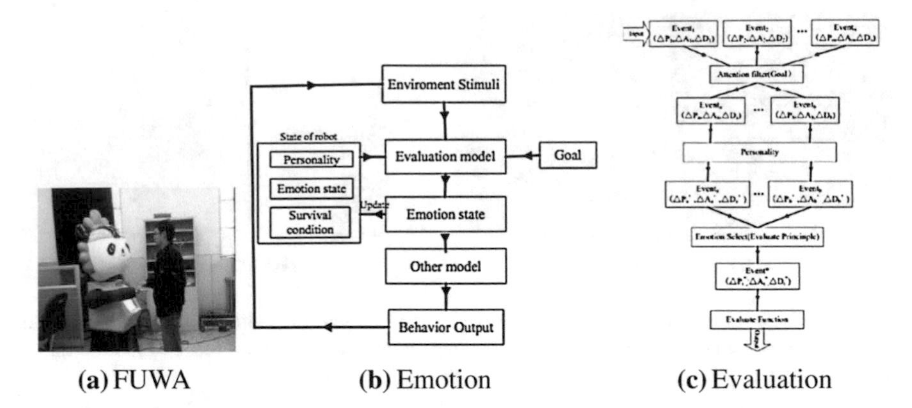

(a) FUWA (b) Emotion (c) Evaluation

Fig. 3.4 FUWA (Fundan Intelligent Robot service developed at the Fudan University in Shanghai (China): **a** the robot shaking hands with human, **b** the frame of emotion generation, where event perceptions are sent to an evaluation model, in which the state of robot's body and goals are taking into account at the same time, and **c** inspired in OCC model by Ortony et al. [199], the extern stimuli are classified into several events according to some defined features. *Source of figures* Qingji et al. [213]

assessment of how emotion changes through the interaction with external stimuli, and the influence of inner personality factors (see Mehrabian [166] for the analysis of personality factors in terms of the *PAD* temperament model).

Generally speaking, the increasing interest in robotics—and specifically social robotics—is being a reason for a continuous challenge within the framework of emotion science. This way, the incorporation of human capabilities together with emotion, making it a permanent part of the structure of decision making, communication or behavior, is becoming a scenario to develop a lot of new solutions. The video games industry and the entertainment industry, as well as the robotics industry, have introduced an essential field into the portrait of emotion research. Virtual agents development (and frameworks for these virtual agents) are imposing new requirements regarding intelligence and emotion design.

3.4.2 Emotion and Cognitive Science

Early in 1984, *Braitenberg vehicles* [30] introduced relevant discussions about whether they might be considered as intelligent or even emotional systems. They showed single behaviors supported by a single model that, however, were clearly driven to some type of "own state" improvement. Nowadays, *IBM's Watson Computer* is likely one of the paradigms in cognitive science research. It is a result of decades of research in the fields of natural language processing and artificial intelligence, and showed amazing capabilities of smart understanding at the TV quiz show "Jeopardy!". *Watson* was implemented by using rules related to the environment of the TV show, and models to obtain information from the internet. It currently constitutes an open field of research from which new capabilities regarding intelligent systems are expected to be achieved.

Any system capable of integrating information and developing some type of feedback, is able to update its state and deploy behaviors within the range of complexity that it implements. However, the increasing demand for intelligence and autonomy in systems is calling for improvements, not only about how the system processes information, but about how it interacts with its environment as well. This enquiry asks for models that allow systems to acquire semantics about itself, the environment and their interaction. Systems are usually assumed as *agents* that are coupled with the environment, and that integrate the model which gives the necessary knowledge for controlling the interaction [223]. *Cognitive Architectures* are models that devise a representation of the human reasoning [223], and that represent nowadays a relevant approach towards intelligence and cognition in systems.

I. Cognitive Architecture Research

Cognition is commonly understood as a process of acquiring knowledge through senses, experience and thought. Within the field of artificial systems and robotics, *cognition* concerns intelligent behavior and interchange of information with the environment. To this end, there exist architectures in order to allow processes of learning, reasoning and managing complex goals within their environments.

The field of cognitive research focuses on different purposes, and conforms a complex integration of disciplines such as psychology, artificial intelligence, philosophy and neuroscience. One of the main paradigms in cognitive science is the design of cognitive architectures. That is, technical blueprints to endow systems with fundamental cognitive processing. Usually, this fundamental cognitive processing has common focus of interest and concentrate approaches around models of general intelligence, human-like cognition or specific features of intelligence. The work of Muller [178] describes a relevant review of cognitive architectures, and shows a pertinent portrait of some essential approaches for intelligent agents [178].

A general view among these common focus, shows relevant approaches that display features such as reasoning about the environment (see Laird et al. [131] regarding *SOAR*—Fig. 3.5), symbolic representation Fig. 3.6 (see Anderson et al. [3] and Taatgen [256] regarding *ACT-R architecture*—Fig. 3.7), human-level general intelligence (see Goertzel and Wang [89] regarding *CogPrime architecture*), integrating perception, action and cognition (see Langley et al. [134] regarding *ICARUS*), non-axiomatic reasoning systems (see Wang [266] regarding *NARS architecture*), cognitive architecture conceptual models for machine consciousness (see Franklin et al. [78] regarding *LIDA architecture*—Fig. 3.6) or episodic memory reconstruction (see Brom et al. [34] regarding *POGAMUT architecture*).

The objective of intelligent control has influenced relevant approaches with successful applications Fig. 3.8 that provide systems with capabilities such as improved perception (see Albus and Barbera [2] regarding *RCS architecture*—Fig. 3.9), controllers on the basis of layers with increasing levels of competence (see Brooks et al. [35] regarding *Subsumption architecture*—Fig. 3.8), capability for attention (see

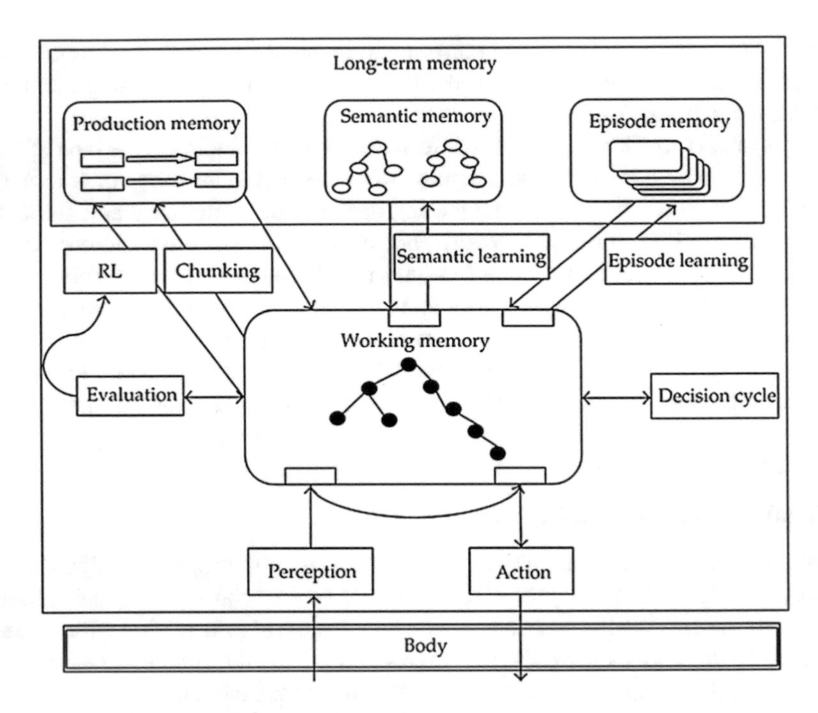

Fig. 3.5 General view of SOAR cognitive architecture

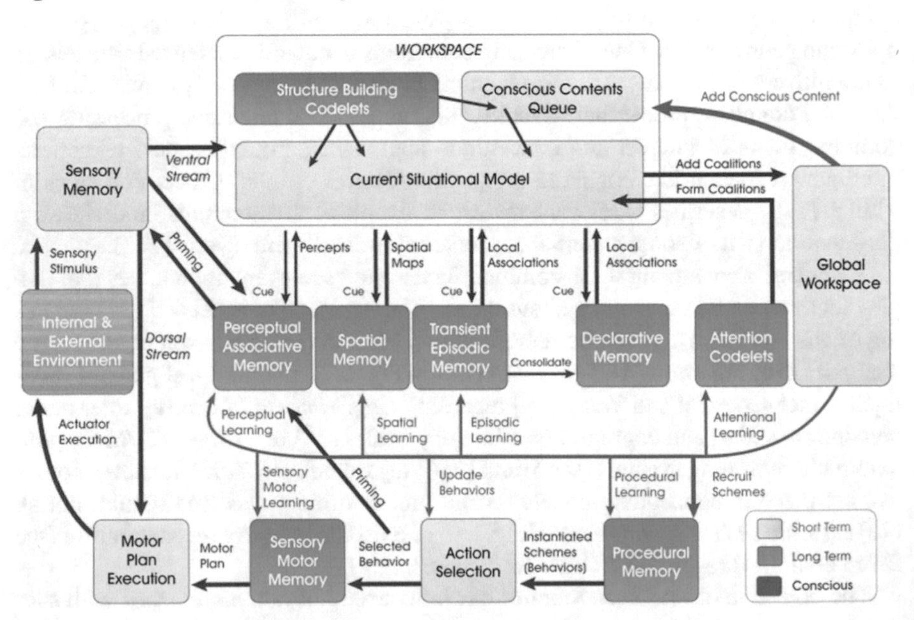

Fig. 3.6 General view of LIDA Cognitive Cycle under the view of Madl et al. [156]. LIDA (Learning Intelligent Distribution Agent) is a later development of IDA [78] with learning integration

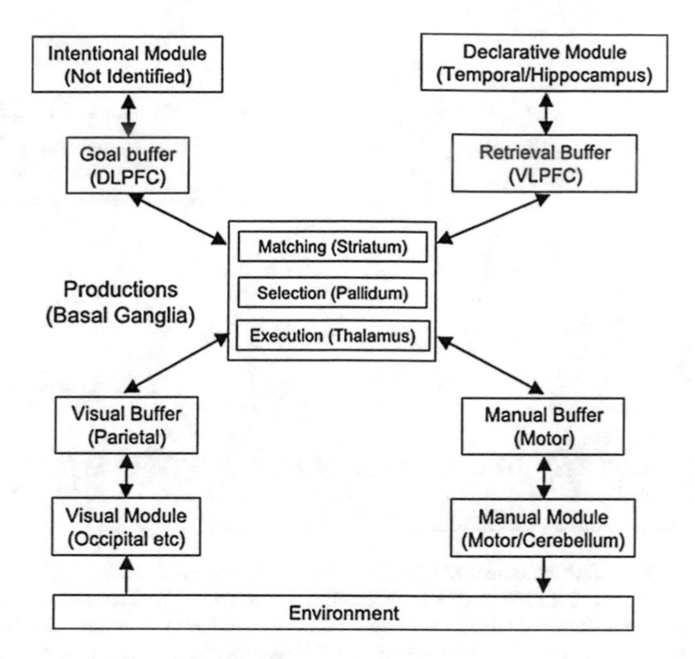

Fig. 3.7 The ACT-R 5.0 model. *Source* [46]

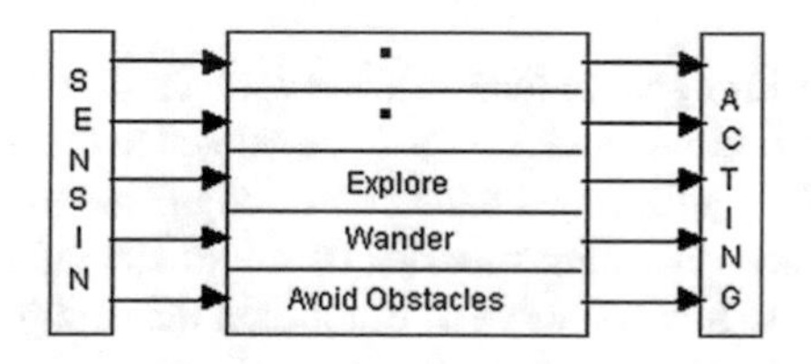

Fig. 3.8 Subsumption Architecture by Brooks et al. [35], used to implement physical robots and purely reactive software agents. The architecture is made up of modules placed in layers, together with augmented finite state machines. Figure of Brook's Subsumption Architecture as shown in [188]

Moreno et al. [176] regarding *CERA CRANIUM architecture*, and Vashist and Loeb [261] regarding *NEXTING architecture*) or representational structures for implicit and explicit cognition (see Sun [250–252], regarding *CLARION architecture*) among the most relevant.

And the objective of learning has led to a wide amount of approaches such as models for error driven learning (see O'Reilly [197] regarding *Leabra architecture*), or a brain-emulating cognition and control architecture for reinforcement learning (see Rohrer [219] regarding *BECCA architecture*).[5]

[5]Grossberg [97] makes a relevant comparison between several network models of cognitive processing and competitive learning.

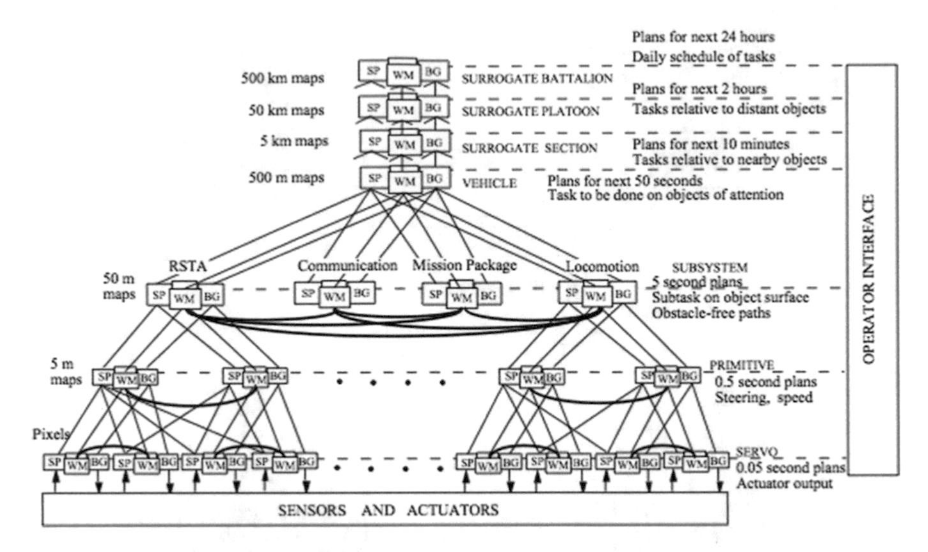

Fig. 3.9 4D/RCS reference model architecture: Real-time control system (RCS) is a cognitive architecture designed to enable any level of intelligent behavior. The first version of RCS was developed as an sensory-interactive goal-directed controller. It has evolved into 4D/RCS over the years, an intelligent controller for industrial robots and several other intelligent tools

The new challenge in cognitive research is the integration of models for resourceful and autonomous thinking, and it seems to be related to the integration of emotion based reasoning [172]. Untill now, I have shown cognitive architectures that implement pertinent solutions regarding theories of mind. Hereafter, I will describe the state of the art related to this objective of emotional reasoning.

II. Emotional Architecture Research

The model of Minsky [171] describes the *Emotion Machine Architecture*, a blueprint that structures those processes that produce commonsense thinking, by organizing the process of thought in reflective layers [171]. This work holds that "emotional states are not especially different from the processes that we call 'thinking'; instead, emotions are certain ways to think that we use to increase our resourcefulness", using the term *resourcefulness* to refer to *intelligence*. The main argument of Minsky [172] is the design of machines with "sufficient diversity" as to describe *resourcefulness* under a "colossal collection of different ways to deal with different situations and predicaments", and no by stemming from a central core (which he refers to as "spirit or self").

However, the most common explanations concern the question of how to build *valuable meaning* related to relevant events. The generalized idea regarding a functional integration of emotion and cognition has resulted in works that analyze the integration of emotional capabilities into cognitive architectures. Commonly, solutions aim at the selection of relevant objectives for the system. To give an example, Newell's theory of mind [182] supports the *SOAR architecture* [131] with the objective of building intelligent agents. Newell [183] revealed in later works an earlier awareness that emotions should be part of a general theory of mind, holding that "we take emotion and affect to be in part like multiple specific goals" [183]. Later extensions of the *SOAR architecture* integrated several modules for non-symbolic processing such as the *appraisal detector* supported by the theory of Scherer [236]. The main objective was the coordination of processes by means of which eliciting inner mechanisms of reward, in order to influence learning and decision making tasks.

Emile is a model of emotional reasoning with explicit planning, focusing on the generalization of some stages within the reasoning process (see Gratch [93]). Gratch [93] holds that *Emile* is able to identify those emotions which emerge from the source of task fulfillment and related to the achievement of the task. However, this model does not successfully explain emotions related to relationships or communication between agents. The *CogAff project* establishes a framework of components (i.e. sub-architectures) as a federated work between three levels of operation (perception, processing core and motor operation) and three levels of processing (reactive, deliberative and reflective or meta processing) (see Sloman [240]). This project aims the analysis of the most common facets of mind, among which there are those referred to as *affective states* that are defined by different types of information processing. The already introduced hybrid architecture *CLARION* [251], implements a dual representational structure that codifies explicit and implicit knowledge. However, this architecture integrates from its earlier versions a motivational system, aimed to guide the interaction of agent and environment by means of computational drives that characterize this motivation. Similarly, the *LIDA architecture* [89] implements a cognitive range from perception to high level reasoning trough three essential stages built within the work of a cognitive cycle: the stages of understanding, awareness and a final phase of action selection and learning. *LIDA* puts into effect the elements of *emotion* and *feeling* as additional data (i.e. informational components), and not as an independent module for emotional operation within the general model of the architecture. This emotional-like data provides knowledge that regards the value that an action taking might have for the system depending on its current state. This way, the system is provided with ability to assess decision making under dangerous circumstances. And *SHAME* is a hybrid neural network architecture that implements previously defined factors for personality. This architecture models the emotional phenomena—concretely emotion learning—in agents within the environment of a virtual world, by creating emotional impulses by means of scalar vectors (see Poel et al. [209]—Fig. 3.10).

Quite different is the challenge of the *MicroPsi project*, that aims at the emergence of emotions from the source of abstract states related to the system operation. That is, the main objective of this architecture is not the simulation of emotions, but the creation of essential grounds for the emergence of emotion (see Bach [13]—Fig. 3.11). This architecture describes the interaction of cognition, emotion and motivation of fixed agents based on the *Psi theory of Dietrich Dorner* (see Bartl-Storck and Dorner [15]).

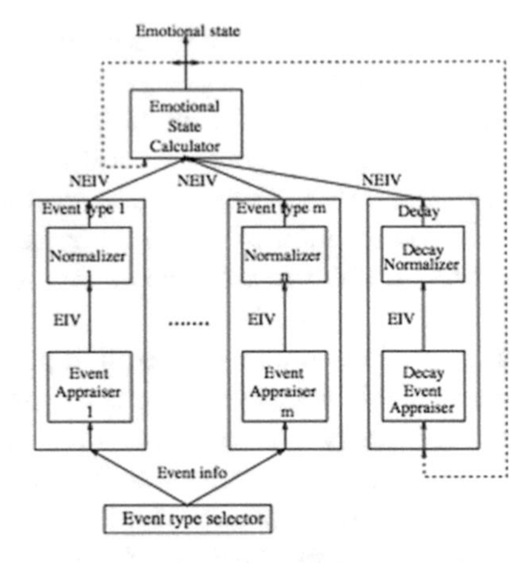

Fig. 3.10 General overview of SHAME architecture: an event appraiser compute an emotion impulse vector (EIV) which is normalized to an emotion vector (NEIV) that acts as an impulser. *Source* [209]

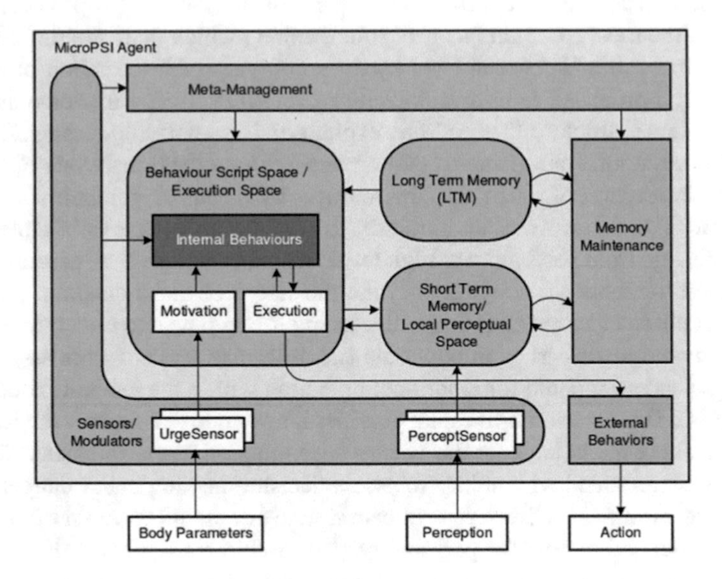

Fig. 3.11 General overview of MicroPSI architecture

3.5 Emotion Modeling

Currently, emotion modeling is a particular sphere of emotion science that aims not only at the design of artificial systems, but also at the definition of a general theory that might widely explain the real phenomena. Rather than present a unique portrait in order to organize theories, this thesis shows a state of the art by means of multiple views. They are those different perspectives from which emotion has been studied, each perspective pertaining to a different distribution, among those theoretical principles followed by each approach. Each theory may use a different principle to reason about the same feature of emotion. And the same theory may be used to obtain different solutions in artificial agents. Therefore, the challenge of making a clear portrait of the state of the art in emotion science, is as complex as the own field in the search for a universal solution.

Formerly, this chapter has showed the study of emotion from the perspective of neuroscience, psychology and philosophy. Subsequent to this description, this thesis makes a representation of how artificial systems have implemented emotion to their advantage, in order to improve intelligence and behavior. And how each particular perspective in engineering sciences, has provided different paths towards emotion-driven technology. However, as the last in the series of related approaches herein showed, a final reasoning regarding emotional models and their relationship with the most common emotional theories is requiered in order to put an end to this analysis.

I. Comments on objectives in emotional modeling The more the complexity of modern technical systems grows, the closer it gets to the complexity of biological agents. Engineers look for solutions in biological systems trying to find patterns in their designs to solve problems in artificial systems. However, distinguishing different parts in emotional research is essential in order to understand the usability of results.

From an engineering perspective, there are strong differences between the *emotion-recognition* research, the *emotion-expression* research and *emotion-feeling* research. Even if they are part of the same biological functionality, they are three discrete parts that currently are being used as individuals in human-made systems.

1. *Emotion Recognition*
   It is used in a wide spectrum of applications with no need to discover whether the artificial system that recognizes the emotion, feels the emotion. It is used just as a means to capture new information from the basis of the explicit information. For example, sometimes the market is interested in learning how people would behave with respect to some issue, in order to trigger new lines of solutions. Hence, with this purpose in mind, the emotional content that emerges from the flows of data in social networks would be evaluated. Other times, the intent is to obtain an analogous explanation from the observed patterns of expressions or behaviors.
2. *Emotion Expression*
   It is used in a different spectrum of applications as a means to influence the environment of the system, in order to obtain something from this environment. As an example of this: depending on the applications, human–machine interaction might require that humans have a better understanding of the machine's messages. In order to achieve this, machines are prepared to trigger emotional messages. That is, engineers use the knowledge about our emotional communication, and they deploy machines that trigger copies of these expressions together with the message. It allows for a better understanding of the machine's message and it is a key result in some human–machine interaction environments.

3. *Emotion and Feeling*
   This is a different problem. How an emotion emerges within an artificial system, how it is used by the system or even what is the artificial emotion, are conforming a wide field of multidisciplinary research full of questions still without a complete explanation. This work suggests that emotion endows the system with some sort of explanation of its environment in the form of value. This value is not explicit information that comes from the environment, but emerging information that explains how the environment is influencing the system concerns. In the end, it is solving issues about how the system should use the set of its available resources (the ones it has for its behaviour) concerning its inner drives (from which the system analyzes how it should be desirable to behave).

II. Relevant works on emotion modeling In the end, emotion is a complex phenomenon with subtle interactions among subjective and objective factors [126]. However, besides its multidimensional character (sometimes even multidimensionality related to the emotional intensity as Frijda et al. [82] holds [82]), there is not much debate about considering emotion as a feature of live systems for adapting to their environment throughout value-based semantics. From the perspective of cognitive systems, biological emotion constitutes the strategy to configure value-based systems by using self representation of the agent state (see Damasio [54] and Picard [206]). The cognitive perspective of emotion suggests that emotion is the result of patterns of evaluation of relevant stimuli (see Sanz et al. [231]) which are carried out by means of some essential bodily processes.

Emotion science delineates some essential processes as part of the phenomena nature. *Homeostasis* describes mechanisms that hold constant a controlled variable, by means of sensing its deviation from a set point, and feeding-back to amend the error. The concept of *allostasis* refers to the ability to maintain stability through goal directed changes [247], describing mechanisms that override local feedback to meet anticipated demands. The idea of '*stability through change*' refers to physiological changes that anticipate requirements on the basis of local feedback. The emotion concept of action readiness captures this idea of change to be prepared for action. And finally, *arousal* is understood as the regulation strategy of activation while *appraisal* refers the regulation strategy of interpreting the meaning of a stimulus [204].

EMA is a computational framework that illustrates how emotion has been applied to the problem of modeling human emotional behavior in a virtual system (see Gratch and Marsella [94]—Fig. 3.12). This framework is on the ground of two essential facets in emotion science, i.e. a computational model of appraisal and coping Gratch and Marsella [94]. A formal application of this model was made, in order to analyze emotional state of simulated agents, while confronting humans and scenarios in a virtual military surrounding (see *Mission Rehearsal Exercise (MRE)*.

Relevant theories of appraisal (such as those of Ortony et al. [199], Fridja et al. [79], Scherer et al. [237] or Lazarus [137]) have resulted in models that implement solutions on the basis of their proposed principles. Earlier works of Fridja et al. [80] have resulted in models such as *TABASCO* (see Staller and Petta [246]—Fig. 3.13), an architecture for software agents aimed at integrating results from functional theories

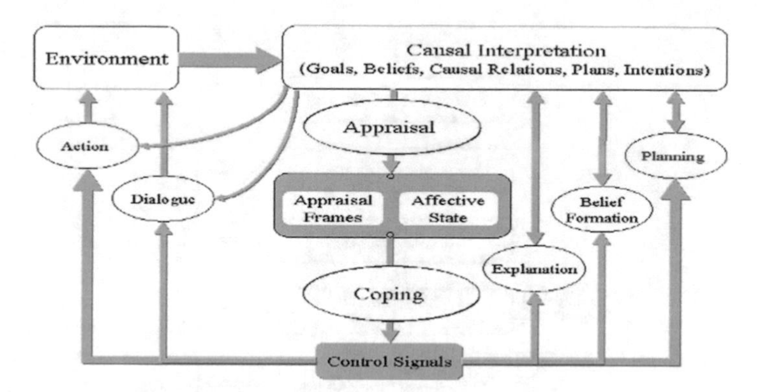

Fig. 3.12 General overview of a computational instantiation of the cognitive-motivational-emotive system for EMA made by Gratch and Marsella [94]. EMA was named in honor of the book of [243] which was titlted 'Emotion and Adaptation'. This work proposed a unified view of appraisal and coping. *Source* Gratch and Marsella [94]

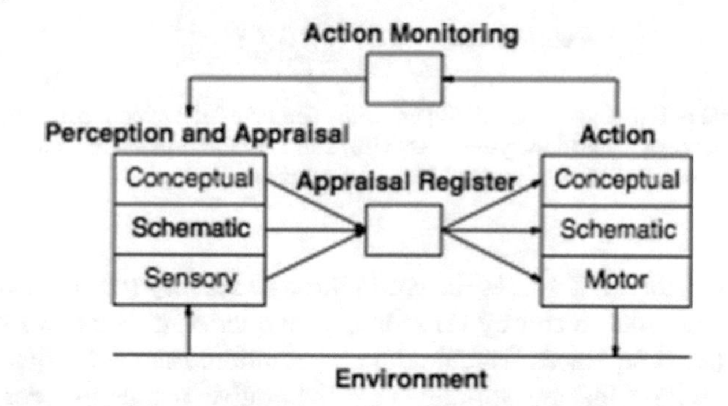

Fig. 3.13 General overview of TABASCO: the goal is the integration of the emotion process within the architecture of an agent, and model emotions as adaptive processes related to the agent-environment interaction. TABASCO's essential idea is that conceptual processing is not the only responsible of sensory-motor, schematic and conceptual processing. It is also affected by the generation of action, as it was proposed by Leventhal [146] in his theory 'perceptual–motor theory of emotion.' *Source* Staller and Petta [246]

in emotion science. The operation of this architecture is related to *ActAffAct*, an environment test-bench with components that help display emotional actions (see Rank [214]).

The work of Frijda1 et al. [80] resulted in some other models such as the *ACRES model* aimed to model emotional processes (see Swagerman [253] and Swagerman [83]), which was refined in a later work to solve issues regarding incoherences in emotional response (see *WILL architecture*, Moffat and Frijda [173]). Figure 3.14 shows the model of emotion proposed by Frijda [80].

As a result of the research made within *The Oz project at Carnegie Mellon* [17] for the simulation of "artistically effective simulated worlds", *TOK* was built (see Bates et al.[18]), a model of agent architecture that supports emotion, goals, reactivity

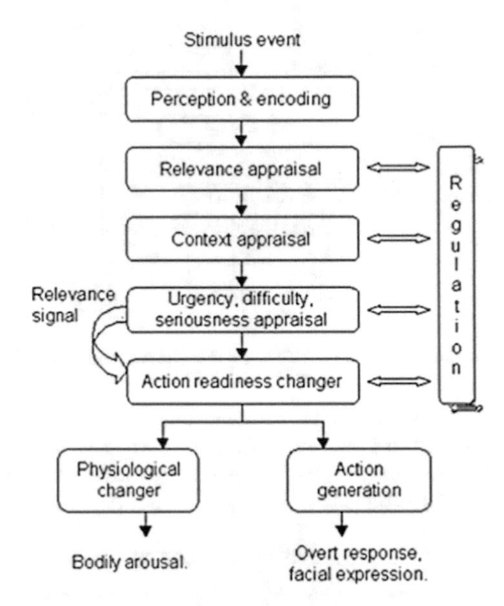

Fig. 3.14 Model of Emotions of Frijda [80] under the central notion of concerns. The two architectures artificial concern realisation system (ACRES) and the later refinement named Acres WILL, began as an implementation of Frijda [80] theory of emotions

and social behavior. *TOK* senses the world through sensory routines and integrated sense models in order to convey sense language queries to other two models (*EM architecture* [217] for standards, attitudes and emotions, and *HAP engine* [153] for goals and behaviors), that are working in a collaborative manner in order to deal with behavior features and goal success.

The *OCC model* of emotions by Ortony et al. [199] combined with the *Five Factor Model* of personality (see McCrae and John [161]) and the *model of moods* proposed by Mehrabian [167] resulted in the *ALMA model* (see Fig. 3.15). It is a

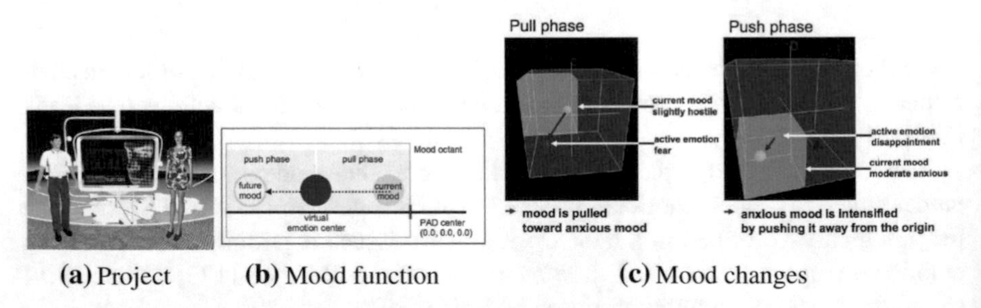

Fig. 3.15 ALMA is part of the VirtualHuman Project to develop interactive characters with conversational skills. ALMA provides a personality profile, real-time emotions and moods: **a** screenshot of the VirtualHuman system, **b** pull and push mood change function (*Source of figures* **a** and **b** Gebhard [87]), and **c** Mood changes in ALMA. *Source of figure*: https://www.dfki.de

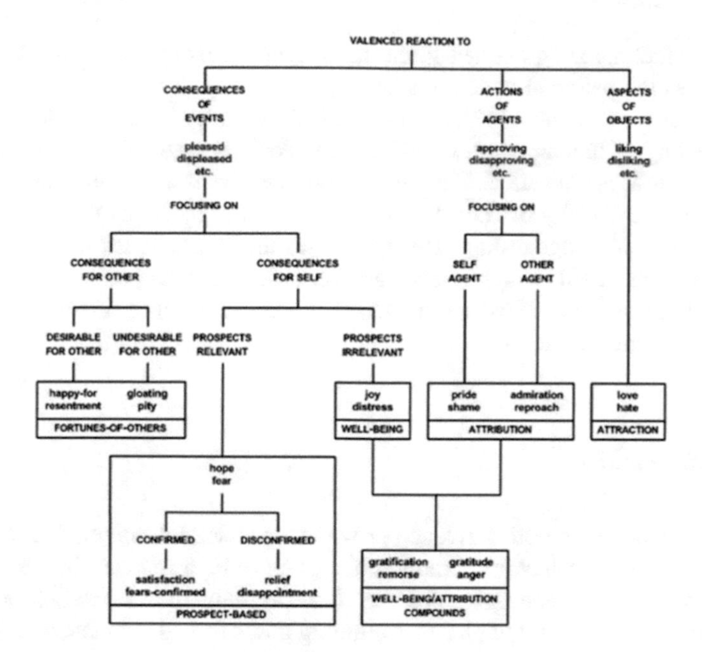

Fig. 3.16 Global structure of emotion types in OCC by Ortony et al. [199]

layered model of affect that integrates "three major affective characteristics: emotions, moods and personality" in order to cover "short, medium and long term affect" (see Gebhard [87]).

The *OCC model* of [199] (see Fig. 3.16) is also an essential support for the *Affective Reasoner model (AR)* of Elliott [70]. The *AR model* focuses on environments with multiple agents in order to model emotions among these agents under the premise that "emotions have much to do with intelligent reasoning". This work proposes a simulation environment with the argument that reasoning is better explored by "simulating a world and populating it with agents capable of participating in emotional episodes".

There is a relevant trend related to the study of the dynamic of change in emotion, in order to model alterations or modifications within the systems through interaction and over time. This is a useful way of identifying critical features of the emotional responses, in order to align the inner processes of the system with the strategies of their missions (allowing this way for some type of self organization according to the criteria of wellness). This way, several computational models of emotion consider the dynamic nature of the phenomena such as the *Cathexis model* proposed by Velkquez [262]. It is a distributed computational model which aims to represent the dynamic nature of distinct facets of the emotional phenomena, inspired by a multidisciplinary work in several fields such as ethology, neurobiology and psychology. This model devised the representation of emotion generation, as well as several other models for motivation and action-selection. This viewpoint is argued by Lowe and Ziemke [152], who hold: "feelings are grounded upon neural-dynamic representations (elevated and stable activation patterns) of action tendency".

As argued by Canamero [42], emotion modeling can even pave a path towards the understanding of human emotions [42]. This idea has resulted in empirical works aimed to investigate the effect of a robot emotional expression in humans. From this perspective, the work by Beck et al. [19] focuses on the objective of creating an affect space for body language testing. This work uses the "Circumplex Model of Affect" of Posner et al. [210], which holds that "all affective states arise from cognitive interpretations of core neural sensations that are the product of two independent neurophysiological systems".

3.6 Discussion

It is quite clear that emotion science provides an essential knowledge in order to improve intelligent behaviors. Emotion is argued to be a relevant influence on the assessment of events, conveying abilities to the systems in order to behave accordingly to their essential principles of wellbeing and survival. Somehow, it seems to be the source of some type of autonomy, since the concept of *autonomy* refers to some sort of freedom from external control or influence, i.e. with some type of independence regarding the capacity of an agent to act in accordance with [what might be called] personal goals. From this perspective, an essential goal of life systems is survivability and environmental adaptation—whether this environment is the physical world or the most complex social environment. As it seems that emotion helps to obtain responses in order to optimize personal objectives, it can be thought that—in some sense—a computational model of emotion should provide new scenarios to control uncertainties in systems. And the means seem to be by transferring the aftereffects of this uncertainty to some type of inner domain within the systems, and dealing with uncertainty on the basis of references for wellness.

This work thus focuses on computational models of emotion that might allow for additional improvements on systems. The main objective of this thesis is a first lay out of an emotional model that might inspire future solutions to better manage uncertainty, by connecting the system with its own wellness and principles at runtime. It thus intends to propose a method focused on providing feedback to the system with valuable meaning, in order to reason not only with regard to its adaptivity, but also with regard to dynamical evaluations concerning the inner state of the system.

I have assumed that baselines and foundations of emotion come from a vast amount of sources that provide knowledge about the real phenomena. And from this assumption, I have built a theory by joining several elements of different theories. Consequently, this chapter has been built in reply to the requirement of a portrait of emotion science and the objective of showing the logic that preceeds this research. And leaving the description of logic, principles, theories and foundations that support the proposed solution of this work for the following chapter.

Once described herein an essential portrait of emotion science, the next chapter will describe the range of conclusions approached in this thesis on the basis of the theories that have supported them. Finally, the amount of inferences will be the baselines and principles to build a computational theory, and will end in the approached model of emotion that proposes this work.

Chapter 4
Principles for Computational Emotion

In this chapter and following in this dissertation, we will differentiate *artificial emotion* from *computational emotion* in the sense that the former will refer to the complete phenomena that an artificial system might deploy, and the later will refer to the set of models that might allow for that phenomena.

The major challenge in *artificial emotion* is caused by theoretical confusion. We remark the requirement of thinking hard about the wide range of theories in emotion science in order to understand and explain the computational problem. There is no complete theory from which to build the computational solution, and this is because the meaning of the focus and the problem have always been a puzzle.

Our purpose is the arrangement of common theories in relation to each other according to a particular pattern, which allows for a complete explanation of the emotional problem focusing on artificial systems. The main objective of this chapter is to identify those principles coming from accepted theories of emotion, in order to find principles of design that can be placed appropriately to conform a computational theory.

The most common method used for getting information about the various aspects in emotion science, is to observe those aspects and also the various processes related to them. Hence, it can be argued that observation acts as a fundamental method of getting information about emotion to comprehend how it works.

Observation, however, is not only seeing the phenomenon, but watching it carefully and trying to understand it in depth in order to get some relevant information about it. This is a critical issue in emotion science since the majority of essential processes are inner processes which have an intrinsic nature of subjectivity. Furthermore, some aspects are not visible from a single point of view and the understanding of [observed] data strongly depends on the perspective of the observer as well.

Emotion science is concerned with the very essence of subjectivity and interpretation. Besides this, the variety of frameworks in which emotion is theorized presents a never-ending complexity that influences negatively as well. To complete the state

M.G. Sánchez-Escribano, *Engineering Computational Emotion—A Reference Model for Emotion in Artificial Systems*, Cognitive Systems Monographs 33,
DOI 10.1007/978-3-319-59430-9_4

of the problem, artificial systems still present challenges such as those concerning perception, awareness or consciousness. So to the greatest extent, solutions do not describe an emotion performance analogous to the one that biological systems show.

We propose our thesis from the standpoint of system engineers that focus on improving systems. To broaden the applicability of artificial emotion further than human-like behaviors in human-machine interaction, an interpretation to adequately characterize the computational problem is essential.

As we have previously discussed, one major focus of complexity comes from the multiplicity of frameworks, controversy, terminological constraints as well as non-defined boundaries of the problem. Taking each part from all of these frameworks is required, thereby interpreting them from our perspective, which allows us to make the essential distinctions that concern the problem of computational emotion.

This chapter discusses assumptions and theoretical strategies underpinning the emotion science. Common theories are analyzed and interpreted from a computational perspective in order to describe and interpret emotion from the point of view of an artificial system. The methodology followed has been an interpretative study that has been analyzed through a computational perspective. Our research is based on analyzing assumptions, theories or discussions that might constitute a reasonable explanation in order to develop a computational theory. This computational theory should ensure trustworthiness and appropriate criteria for computational research. This chapter describes major aspects in emotion science to build a complete portrait of the phenomenon, and justifies the essential principles of design.

4.1 Introduction

Following this general introduction about the formats used in this chapter, the following sections are grouped into the following areas in order to describe the complete problem of the emotional phenomenon.

1. *The State of the Challenge*
   This section focuses on a critical view about the challenge of designing computational emotion.
2. *Conceptualization of Artificial Emotion*
   This section focuses on declaring which perspectives have been accepted within the range of emotion science ramifications. The exploration and selection of this perspective has been part of our study and it cannot be underestimated because, once a perspective is chosen, it becomes determinant for later decisions. It will define the approach from which we will describe our thesis, and the differentiation of some essential terms. This section explains the differences among *artificial emotion*, *computational emotion* and *computational feeling*.
3. *Appraisal and Computational Emotion*
   This section presents the foundations for emotional assessment and the conceptualization of *computational emotion* and the *appraisal* as the essence of its

operation. Given an understanding of these theories here described, the principles from which final proposals concerning design will be defined can be devised. These proposals will concern *computational appraisal*, the *computational process of appraisal* and the concept of *measurement of compliance*.

4. *Perception*
   Emotion design is strongly joined to perception, which contains logically related processes to convey qualities to the system. This section identifies the artifacts that are used afterwards to manage perception of emotion. The intent of these artifacts is to decouple the problem from the system design.
5. *Autobiographical Self*
   Memories (i.e. the storage of autobiographical knowledge) may be implemented in various ways and under several objectives in artificial systems. Autobiographical knowledge is conceived in emotion science as an essential part to conform a ground for the *self* and consequently for *emotion*. It seems to have intimate knowledge of personal resources and experiences that influences the coordination of emotional processes. This section describes main features of this memory in order to describe those features that will affect the emotional process.
6. *Representation and Computational Feeling*
   This section discusses and establishes the essential principles from which principles for *computational feeling* can be conceptualized. This goal, however, is quite difficult to achieve. This section reviews the factors under a well known and relevant theory as an strategy to draw the design.
7. *Architectural Theoretical Foundations*
   This section discusses and establishes design decisions concerning the blueprint of a computational system of emotion. In the end, it is software, and design decisions have to encompass every aspect of the emotional system concerning structure, functional behavior and processes interaction.
8. *Alertness, Arousal and Action Readiness*
   This final section includes some relevant characteristics and qualities of emotion that is required to take into account. This section aims to clarify these concepts, in order to provide the perspective under which this thesis will use them.

Our research is arranged in the broad field of emotion science from the perspective of the engineering. We investigate fundamental processes from biology to build computer models that implement and explain these processes. Our research references are works in the state of the art of emotion science, and they constitute the baselines and principles under which this thesis has been developed.

Certain features of life systems are still without a complete explanation, and our research is completely conditioned by this issue. The scientific community have studied and analyzed different perspectives of the emotional phenomenon isolating effects, and commonly under observation integrating subjectivity within theories.

We have tried to discover connections between those relevant theories that have explained essential parts or elements of the emotional phenomenon. And among the broad range of theories, we have chosen those with explanations that allow for computational models. This thesis proposes a single theory which is composed of several well-coordinated theories, each one describing essential parts of the phenomenon, which might contribute to the basic function of computational emotion.

The chapter follows the described scheme regarding sections. Each section will describe one or several accepted theories in order to explain the phenomenon and that will conform those principles and baselines that will condition the design pattern of the computational analogy. In order to differentiate what is proposed in this thesis from those concepts that have origin in other theories, we have framed our proposal under a gray colored form.

4.2 The State of the Challenge

As previously mentioned (see State of the Art), major challenges in artificial emotion regard what, why and how. Commonly, the challenge of computational emotion has boiled down to an increase of fixed evaluations among stimuli, goals (whatever the abstract idea), values or behaviors. It seems as if by copying the biological function we might endow artificial systems with the same final capabilities. This way, the most amazing solutions show systems that behave emotionally regarding our perspective of [emotional] observers, disregarding the fact that we understand emotion because we deploy emotion. It is our emotional system the one that is perceiving emotion, and it is not the system which is actually feeling the emotion. The benefits of these solutions are driven to our interaction with the artificial system.

Quite often happens that solutions stray or follow isolated questions about the real emotion. But we all realize constraints in these solutions; even more, these constraints are well assumed and defined in each work. They are relevant solutions regarding the framework where they are deployed, but not necessarily practical in control design focusing on improving adaptiveness, survivability or robustness.

I. The purpose of analogy Oftentimes there is a lack of perspective regarding requirements in emotion design. Clearly the structure, purpose and operation of man-made systems vary from the structure, purpose and operation of biological systems. According to this, the essential facet should be the exploration of what emotion solves inside of biological agents, to afterwards search for relevant analogies in artificial systems. Since emotion depends on the system in which it operates, a model of emotion should not be built as a direct correlation of the biological feature. It moves the design towards some sort of absurd idea that artificial systems requirements are exactly the same as those of biological systems. Since they are far different systems, their needs might also be extremely different even when the result might influence the same type of responses (intelligent decision-making, behavior, motivations, etc).

To give an example: When washing machines were designed, nobody thought of replicating a human doing the washing work in the way it had been done till that time. It was better to analyze what the effects of the human work on the clothes were, i.e. the removal of dirt. And the final model was the replication of the human movement, not the whole system of a human washing the clothes.

Analogously, our objective should not be the direct implementation from biology (either a mirror-image or a direct matching), but an adapted equivalence of the emotional operation for the real needs of artificial systems. As an utter insanity, sometimes even physiological processes such as those that implement hormonal activity in life systems are replicated. The replication of the hormonal functionality is far from implementing an analogy concerning its causality within the system (since artificial systems have their own artifacts to solve communications and data transference). We hold that artificial emotion (i.e. a computational model of emotion) might not be a direct replication of the biological relationship between emotion and what causes the emotion.

Nevertheless, under a successful design, it should be useful for purposes of explanation by means of inverse engineering. That is, the explanation of some sectors of the real phenomenon on the basis of the theory developed would be desirable. And this essentially requires a new perspective.

II. Direct cause–effect Besides this problem, we should simplify and organize knowledge as well as avoid false beliefs. For example, the relationship between the effect of the environment and the emotional behavior is not as simple as it appears in common solutions. It is more complex than simple cause–effect. The emotional behavior stems from the fact that the essential goals of survival raise an unstable state. It is the environment that makes the goals unstable, but it is the state of these objectives which triggers the emotional processes. They are two different issues. Under this chain of misconceptions, designers usually decide when an event should trigger processes that become emotion for the system. By assuming a direct cause–effect, solutions will deploy similar or modulated behavior related to a concrete stimuli (even when life systems empirically show deviations in emotional responses concerning similar stimuli).

In all these solutions, system designers are observers that determine the causality of an emotion, as well as the meaning or consequences of that emotion for the system. Designers are getting rid of the system-assessments to replace these system-assessments by those made by them. And results cannot be far from the biological benefits of emotion; the real influence of emotion over the system should come from the inner perspective of the system about its own emotional awareness. An emotional observer may draw conclusions about the observed emotion, but not decide about the emotion of the observed system.

III. Isolated parts of the same phenomenon Emotional implementations are limited to spaces of bounded dimensions (expression, decision-making, motivation, etc.) and they are showing low accuracy and low tolerance to uncertainty outside these bounded environments. The larger part of emotional implementations create new information on the basis of the emotional models for being used by the system and this is not incorrect at all. What may be missing is that we remove part of some essential information in these models from the very initial phase of design.

Largely, models of stimulus-response are built on the basis of models that establish a relation between effects and targets of well-being (concerning some previously defined scenarios). In some way, this is a linear approximation to the emotional function at a given domain (system state, environment, emotional influence, etc.) that gives us the chance of addressing the influence of discrete emotion within local parts of the system. Once the system runs outside predefined parameters by design, the emotional model does not work as expected. Many theories assume models from the basis of isolation of certain functions within these bounded scenarios. However, there are lots of related processes (either to the agent as well as to its environment) that are not modeled but nevertheless they are affecting the processes concerning this—initially isolated—function.

Subscribing the theory of Hayek [105], the effects of stimuli within the system also have semantics for the system, and part of these effects are avoided in models. Later effects that might emerge from former effects, conform an associated semantic which might be part of the pattern of stimuli that feature and influence the emotional response. Partial models avoid essential parts of the system and consequently avoid parts of the phenomenon, interactions, relationships, influences, etc. And solutions become sequential processes of perception–emotion-behavior without much meaning for systems.

4.3 Conceptualization of Artificial Emotion

Emotions make the hard work of processing all the available resources within the biological systems to ensure its own survivability. Actually, to give an accurate argument, the emotion supports the use of system resources to ensure that the system's work for survival takes place (whether the system survives or not). The reasons why emotion is so demanded in artificial systems nowadays have already been discussed. However, emotion science states problems still unsolved. Focusing on design, some of the key challenges are (a) how emotion coordinates system elements, (b) how arrangements are done, and (c) how it coordinates the system activity under the coarse-grain goal of survivability.

Emotion should start those essential processes to enable emotional-like responses. Processes such as attention, motivation, perception, memory, thoughts or communication, which are major challenges in artificial systems. The nature of emotion is intrinsically joined (as a minimum) to the concepts of *goal*, *tendencies*, *thoughts* and *feelings* [221], as well as to additional issues that still remain without agreement.

This section proposes the computational conceptualization of *Artificial Emotion*. For the sake of a successful design, we need to understand from a computational point of view those common concepts in emotion science. Along this section, we will justify the main pillars that support the basis of our theory at the same time that we propose computational perspectives from which the next chapter will be ground. To this end, we will explain the portrait of *Artificial Emotion* by using well known works that describe the emotional process, emotional events, value, meaning, appraisal, the dynamic of the emotion and the emergence of feeling, which we will accept and assume as correct for our purposes.

Thesis Rationale ♣ 1 *Agent-Generic-System (AGSys) and Emotional-System (ESys).* Prior to continuing this section, we will introduce our terminological perspective regarding what we will name *Generic-System (AGSys)* and *Emotional-System (ESys)*. Under the requirement of a general applicability, we argue the requirement of decoupling the emotional domain (i.e. the emotional problem) from any other domain that might affect the system. We will use the term *generic* to denote not specific systems that might house and deploy emotion. That is, we will conceive a *AGSys* as a knowledge-based system capable of deploying sensory-perceptual-conceptual processes, able to represent *abstract objects* from the perspective of Gruber [98]

This separation is essential from the perspective of design, since it allows the differentiation of artifacts that are related to the system from those that are related to the 'emotion' of the system. From our design perspective, we have established and assumed the difference between the system that deploys emotion (i.e. a Agent-Generic-System labeled as *AGSys*), from the system that [innerly inside of *AGSys*] builds that emotion and that we have labeled *ESys*. This perspective is not arguing however that they are decoupled systems regarding the emotional assessment. That is, we are not arguing that they are just two systems whose outputs are influencing one of each other. We rather intuit a federated work of assessment with continuous flows of information from one to each other, in order to allow the recursive phenomenon of appraisal that show life systems in their emotional reality.

4.3.1 The Cognitive Approach of Emotion

Even though there is no agreement on a common definition of *emotion*, it is however accepted that emotion guides our perceptual system by building relationships between the environment and our inner concerns. And these relationships are argued to be symbolically represented in our minds while we think [54, 105, 151].

The nature of such an intricate work of biological emotion is not easy to solve. Emotion is an essential part of cognitive processes and becomes a key element that influences our reasoning:

> (...) it is accepted that our interpretation of our senses – i.e.: the specific cognitive equivalence we establish – may vary according to our emotions, actions, bodies, etc. This leads to embodied cognition, term which stresses the relation between the process and the body in which it is grounded.
>
> López Paniagua [151]

In case computational systems could deploy such a feature, should implement some type of federated work among references, relationships and representations to deploy analogous processes to those of the emotional phenomenon. But not only visible processes regarding an external observer, but inner processes concerning the system.

There is a wide agreement about the dynamic nature of emotion. However, there is a wide discussion regarding other issues such as the categorization of the emotional states [71] or the emergence of emotion as a consequence of evaluations and predictions among others [237]. We will accept emotion as a multi-component system whose parts are intended to achieve functions of different nature (physiological, cognitive, qualitative and subjective responses) in the sense that Scherer et al. [237] and

Leventhal and Scherer [148] explain. Leventhal and Scherer justify their perspective by arguing that "terms emotion and cognition refer to complex, behavioral compounds whose make-up changes over the organism's life span and these behavioral compounds are the product of a changing, multi-component processing system." This work assumes a complex operation of emotion, and maintains that reflexes (i.e. stereotyped response mechanisms) are not equivalent to emotion if they are taken alone [148].

Thesis Rationale ♣ 2 *Emotion as a multi-component system.*
We will accept emotion as a multi-component system whose parts are intended to achieve functions of different nature (physiological, cognitive, qualitative and subjective responses) in the sense that Scherer et al. [237] and Leventhal and Scherer [148] explain.

The vision of Fernández-Abascal et al. [75] is a bit more generalized and describes emotion as a multi-dimensional process aimed to analyze significant events, do subjective analysis, express and communicate, prepare for action, influence behavior and change physiological activity. This viewpoint includes the most common perspectives in emotion study such as emotions and bodily reactions (see [118, 132]), emotional expression (see [67, 116]), emotional communication (c.f. [193]) or value and appraisal (see [199]).

Someway, it is reasonable to accept that these processes of analysis in life systems are cognitive processes of evaluation. And that these processes help systems for referring the perceived environments to their own concerns. We will assume cognitive approach of emotion, and accept the idea of existing [some type of] interconnections among system's concerns and the events coming from the environment through which it moves. Roseman [220] defends the primacy of this *cognitive approach* in producing emotion (rather than behavioral or physiological approaches) and assumes an interpretation of events rather than the fact than events determine the emotion. Oatley and Johnson-Laird aimed a computational theory of emotion but the implementation was not ever made. He stepped up this cognitive approach by holding that emotion is an essential part in the organization of cognitive processes:

> According to cognitive approaches, emotions are important because they relate outer events and other people to inner concerns. A principle of these approaches is that an emotion is a judgment of value (Aristotle Rhetoric, [51]); for instance, that a particular event is important, that it is pleasant to be with a certain person, or that a specific concern is urgent.
>
> Oatley and Johnson-Laird [191]

And Ortony et al. followed a similar line of wok analyzing how cognition contributes to emotion as emotion "involve[s] feelings and experience, they involve physiology and behavior, and they involve cognitions and conceptualizations." [199].

Thesis Rationale ♣ 3 *Cognitive approach of emotion.*
We will assume a cognitive approach of emotion, and accept the idea of existing [some type of] interconnections among system's concerns and the events coming from the environment through which it moves.

By accepting this cognitive argument, we are accepting the cognitive concept of *appraisal* as one of the essentials features of the emotion experience. This term was introduced by Arnold, and conceived this *appraisal* as a progression of events related by the work of perception that are responsible of emotion arising [5]. Under the work of Arnold, primary appraisals start the emotional progression that induces both experience and behavior, holding that physiological changes are part of this progression, they are not the initial point [5].

Thesis Rationale ♣ 4 *Appraisal as an essential feature of emotion.*
We accept the cognitive concept of *appraisal* as one of the essentials features of the emotion experience.

Another debatable question is about the character, computation and quantification of emotion, and what are the essentials under which emotion emerges. This is a matter worded by Fox considering that "an important question concerns whether "cognitive appraisals" should be considered as "causes" of emotions (which is implied by most appraisal-based models—Arnold [5]; Lazarus [135]), or whether they should be thought of as "components" of emotion" [77]. This question will be studied, later on following sections, in further detail.

Additional insights into the problem of *appraisal* regard the discrete versus continuous character of the assessments. Among the works in the state of the art, the one of Scherer et al. assumes a continuous evaluation process, suggesting that the functionally defined subsystems in organisms as well as the components of emotion are multiply and recursively interrelated (that is, changes in any component can imply changes in other components) [237]. Scherer et al. argues a continuous assessment rather than a discrete perspective that establishes the emotion as a direct consequence of *appraisal*:

> In the framework of the component process model, emotion is defined as an episode of interrelated, synchronized changes in the states of all or most of the five organismic subsystems in response to the evaluation of an external or internal stimulus event relevant to major concerns of the organism. In other words, it is suggested to use the term emotion only for those periods of time during which many organismic subsystems are coupled or synchronized to produce an adaptive reaction to an event that is considered as central to the individual's well-being.

Table 4.1 is a compendium of the *Component Process Model* theory showed by Scherer et al., and describes the collection of concise and detailed information about the emotion component, the organismic subsystem related and; the emotion function:

We will assume thus two parts in our discourse regarding the emotional-like assessment, in a joined feature that cannot be independently established as problem. They are the requirement of a multidimensional appraisal together with the feature of a continuous assessment.

Table 4.1 Relationships between the functions and components of emotion and the organismic subsystems that subserve them

Emotion component	Organismic subsystem	Emotion function
Peripheral efference component (cognitive component, appraisal)	Information processing (CNS)	Evaluation of objects and events
Neurophysiological component (bodily symptoms)	Support (CNS, NES, ANS)	System regulation
Motivational component (action tendencies)	Executive (CNS)	Preparation and direction of action
Motor expression component (facial and vocal expression)	Action (SNS)	Communication of reaction and behavioral intention
Subjective feeling component (emotional experience)	Monitor (CNS)	Monitoring of internal state and organism-environment interaction

CNS central nervous system, *NES* neuro-endocrine system, *ANS* autonomic nervous system, *SNS* somatic nervous system. The organismic subsystems are theoretically postulated functional units or networks (adaptation from Scherer et al. [237] and Fox [77])

Thesis Rationale ♣ 5 *Continuous process of assessment*.
We will assume thus two parts in our discourse regarding the emotional-like assessment under the ideas of Scherer et al., a and b, in a joined feature that cannot be independently established being: (a) the idea of a continuous assessment rather than a discrete perspective, which establishes the emotion as a direct consequence of *appraisal*, and (b) the idea of a multidimensional feature of emotion, sometimes close to the ideas of a multidimensional assessment and the evaluation of interoceptive realities.

This imposes the design of a framework where the notions of 'continuous' and 'multi-dimensional' be computationally solvable, as well as our intuition about the design of some associated dynamics that helps in this objective.

4.3.2 Emotion and Abstractions

Once accepted the dimensional and continuous features of the emotional assessment, the upcoming problem (under a computational viewpoint) is the analysis of those relationships that characterize the work of such different realities and that can identify their fundamental continuous work.

We will firstly assume that from the perspective of a computable design, bodily symptoms, subjective feeling (i.e. emotional experience), cognitive process (i.e. appraisal), motor expression or motivation (i.e. action tendencies) among others, they are all components that involve different abstractions. Their realities are different since each one is focused to provide different operations, different abstractions or different types of informational flows. Consequently, they have to be designed by following different principles and requirements.

From a computer science viewpoint, "abstraction is the selection of a set of concepts to represent a more complex whole" as as process that moves upward from details to summarizing concepts Taylor et al. [257, p. 87]. Cloutier et al. argues that "abstraction is a very useful concept in describing complex constructs" and that it might be thought as a continuous sequence of levels in which adjacent layers are not perceptibly different from each other, although the extremes are quite distinct [48, p. 21]. Even when it seems easy to explain, authors hold that the key problem of abstraction resides in implementing these appropriate levels where details of design be in accordance with each essential aspect of the reality.

Even when this specific opinion is expressed under computational terms, *abstraction* is formulated among disciplines arguing common thoughts on the topic. And it becomes an essential concept as a means to draw explanations for complex cognitive processes:

> Designing perceptive, deliberative and complex action functions and systems involves a series of difficulties. In general, abstract concepts to be perceived in the environment may not be sensed directly, but identified and characterized from particular combinations of the sensor values. Deliberative mechanisms at high levels of abstraction may not correspond to mathematical operations and may adopt more general forms such as those in animal systems, difficult to map directly into artificial implementations. Abstract action has the problem of task decomposition among others. (...) Abstraction in artificial systems implies undetermination in all system functions. For system operation, it has to be resolved either as part of the design or dynamically by the system itself
>
> López Paniagua [151]

Newell and Simon is one of the most appropriate authors in artificial intelligence, and a relevant contributor to cognitive science. He (as well as several other authors) argues that there is no intelligence without the ability of managing abstractions (this author identifies abstractions as *symbols*) [185].

Newell and Simon hold that intelligent actions are intrinsically joined to the management of abstract data structures, and he even paraphrases those words of McCulloch who claims: "what is a symbol, that intelligence may use it, and intelligence, that it may use a symbol?" [162].

Thesis Rationale ♣ 6 *Abstractions (Symbols).*
Abstractions (i.e. symbols) are key concepts in emotion science, and they are essential tools for explaining the mechanism of emotion. It seems as if there could be no possibility of deploying computational emotion without the ability of managing symbols.

Emotion and intelligence are joined by nature, and consequently it seems as if there could be no possibility of deploying computational emotion without the ability of managing symbols (i.e. abstractions). Abstractions or symbols are key concepts in emotion science. They are essential tools for explaining the mechanism of emotion, and a sine qua non for the construction of a systematics that could aim computational feeling [54].

4.3.3 Computational-Emotion and Computational-Feeling

From our design perspective, we have established and assumed the difference between the system that deploys emotion (i.e. a Agent-Generic-System labeled as *AGSys*), from the system that [innerly inside of *AGSys*] builds that emotion and that we have labeled *ESys*. This is an essential differentiation for the sake of a good comprehension of our arguments. Additionally and regarding now the *ESys* design, the computational conceptualization of other essential components of the emotional phenomenon is critical as well.

Focusing on the operation of *ESys*, we will assume the integration of several core realities (either they can be assumed as processes, entities, patterns, etc.) that each essay has used under its own objectives of explanation. Common terminology used throughout theories in emotion science draws a mixed map of concepts and meanings with no defined boundaries. Somehow, this terminology respects some general principles of meaning, but each author delineates his or her own distinctions in order to establish specific frameworks for explanation. And this is a critical issue that we need to solve in this current work.

The portrait of a computational emotion moves models from emotion science to computer science disciplines. And a proper correlation between emotion science terminology and computational science artifacts is a critical issue that requires an earlier attention in design time. We highlight once more our intention of engineering computational emotion, assuming the need of computational artifacts that are not present in nature. And the requirement of correlating analogous processes that are present in nature and that however do not need to be present in artificial systems in the same form. Herein we will refer to two essential concepts in emotion science, *emotion* and *feeling*.

Relevant essays reveal differences regarding conceptualizations and mutual influences between *emotion* and *feeling* that need to be studied carefully. We will accept the differentiation between computational emotion and feeling, in order to (a) preserve the common vision in emotion science, and (b) differentiate artifacts that help to build the computational solution.

LeDoux differentiates *emotion* and *feeling* by arguing that "contrary to popular belief, conscious feelings are not required to produce emotional responses, which, like cognitive processes, involve unconscious processing mechanisms" [143]. And he reinforced this theory in later works by holding that emotion is not just a conscious feeling:

> Feelings of fear, for example, occur as part of the overall reaction to danger and are no more or less central to the reaction than the behavioral and physiological responses that also occur, such as trembling, running away, sweating, and heart palpitations. What we need to elucidate is not so much the conscious state of fear or the accompanying responses, but the system that detects the danger in the first place. Fear feelings and pounding hearts are both effects caused by the activity of this system, which does its job unconsciously – literally, before we actually know we are in danger. The system that detects danger is the fundamental mechanism of fear, and the behavioral, physiological, and conscious manifestations are the surface responses it orchestrates. This is not meant to imply that feelings are unimportant. It means that if we want to understand feelings we have to dig deeper.
>
> LeDoux [140]

Damasio also agrees this difference, and holds that *emotion* appears just in case that a *feeling self* is created in our minds: "we know that we have an emotion when the sense of a feeling self is created in our minds. Until there is the sense of a feeling self, in both evolutionary terms as well as in developing individual, there exist well-orchestated responses, which constitute an emotion, and ensuring brain representations, which constitute a feeling. But we only know that we feel an emotion when we sense that emotion is sensed as happening in our organism." [54].

Thesis Rationale ♣ 7 *Artificial Emotion.*
In order to prevent ambiguities in our explanation, we will consider *artificial emotion* as the complete phenomena realized in life systems. That is, *artificial emotion* will be conceived as the whole state of the system deriving from its own circumstances concerning wellness.

From an artificial emotion perspective, the computational problem regards the isolation and the course from *emotion* to *feeling*. This problem raises key issues about how artificial emotion can be conceptualized as a whole. Kleinginna and Kleinginna [126] define emotions as complex interactions between subjective and objective factors (together with neural and hormonal processes as means) that become experiences, cognitive processes such as perceptual effects, physiological changes for adaptiveness and situational adaptiveness (a.k.a. readiness) and expressions. This allows us to understand artificial emotion on the basis of two domains:(a) a functional operation within the inner domain of the system (complex interactions between subjective and objective factors), and (b) functional operations that concerns the behavior of the system (complex interactions with the environment such as readiness or expressions).

Under the common understanding of the emotional phenomena, the two domains are essential in order to complete the emotional functionality. However, the issue that concerns us now is that of the inner domain operation of the system, since it seems to be responsible of the latter.

Computational operation requires an essential differentiation between what is named *emotion* and the processes that cause emotion, and what is named *feeling* and the processes that cause feeling. Emotional phenomenon is sustained by a mess of cues, signals, processes, etc. that cooperate to assess and challenge wellness. That is, there is a coordinative work among several non-visible parts that serves to the system in order to deploy the phenomenon. This differentiation is supported by the idea of LeDoux that holds: "emotional feelings and emotional responses are effects caused by the activity of a common underlying system". Under this rationale, we computationally conceive differences between the results and the processes that cause these results.

Thesis Rationale ♣ 8 *Computational Process of Emotion (cp-Emotion).*
We conceive cp-Emotion as the whole set of processes and means required to conform the emotional assessment, and that allow for emotion. Herein we are referring to emotion in relation to the concepts of 'what happens' in Damasio [54], and 'emotion' in LeDoux [140] and Izard [117].

There is a wide agreement in conceiving symbolic representations of the emotion dynamics as *feelings* accepting in this way the theory of Damasio [54]. That is, mental representations of the physiological changes occurring during the emotion. So, analogously to our previous proposal, we will conceive also in this case a separation between the phenomenon and the processes that support the phenomenon.

Thesis Rationale ♣ 9 *Computational Process of Feeling (cp-Feeling).*
We conceive cp-Feeling as the whole set of processes and means required to build the *feeling*, and that allow for the emotional experience. Herein, we are referring to feeling in relation to the concepts of "the feeling of what happens" in Damasio [54], "feeling" in LeDoux [140] and "emotion schema" in Izard [117].

Going ahead in our explanation, we argue that *emotion* and *feeling* cannot be conceived as identical computational viewpoints since they are referring to different domains and realities. As we have previously explained, abstraction design strongly conditions this issue because computational systems require the identification of bounded functions. The computational realization of the *emotion* and *feeling* in a way that changes occurring during *emotion* can be integrated in the construction of *feeling* is thus essential. By conceiving *reality* as the space where a punctual abstraction is defined, we can argue that they are two realities of the same problem that require to be computationally integrated. Thus, we propose to conceptualize a midway step that acts as an intermediate stage between these two realities.

Thesis Rationale ♣ 10 *valuable-State (v-State).*
We propose to conceptualize a midway step that acts as an intermediate stage between these two realities: an stage that keeps the result of the prime functional perspective in the form of value, and that transfers this value in order to be used for the later functional perspective. This conceptualization of valuable-State is not intended here to maintain the result of the emotional evaluation, but just the middle age between the *emotion* and *feeling*. It is not thus the concept of *valence* as it is used in emotion science.

This *v-State* is conceived as a middle stage between *cp-Emotion* and *cp-Feeling*. The process *cp-Emotion* results—among other elements—in a *c-Valence* concerning the event, and this *c-Valence* will be used within two essential processes: (a) the processes for immediate responses (they will be described later in this chapter), and (b) the process of *cp-Feeling*. However, *cp-Feeling* needs more than just an assessment that results in a *c-Valence*. It requires additional information such as the knowledge about the event and the state of activation of the system. The whole information is thus conceived to be integrated within this *v-State* in order to enable a connection between two processes that works at different levels of abstraction.

4.4 Appraisal and Computational Emotion

Generally speaking, *appraisal* refers to assessments on the basis of stimulus perception. Also generic but more detailed definitions, regard cognitive evaluations and interpretations of the phenomena or the events that concern the system.[1] More

[1] http://psychologydictionary.org/.

	Frijda (1986)	Roseman (1984)	Scherer (1984)	Smith/Ellsworth (1985)
	Change		Novelty	Attentional activity
			suddenness	
Novelty	Familiarity		familiarity	
Valence	Valence		Intrinsic pleasantness	Pleasantness
	Focality	Appetitive/aversive motives	Goal significance	Importance
			concern relevance	
Goals/needs	Certainty	Certainty	outcome probability	Certainty
Agency	Intent/Self-other	Agency	cause: agent	Human agency
			cause: motive	
Norms/values	Value relevance		Compatibility with standards	Legitimacy
			external	
			internal	

Fig. 4.1 Comparative overview of major appraisal dimensions as postulated by different theorists from the source of Davidson et al. [61]

often than not, cognitive appraisals are accepted as essential parts of the emotional experience. However, the multiplicity of explanations given to describe the appraisal processes, hinder the conceptualization of the term in a computational way (see Fig. 4.1). Oftentimes, the abstract idea of *appraisal* regards additional abstractions (such as an *interoceptive processing*, *awareness*, *subjective evaluation*, etc.), and the problem of conceptualizing these term grows in complexity. Maybe, a deep understanding on how appraisal works on individuals is essential in order to discover analogies regarding artificial systems.

Foundations concerning emotional evaluation are full of comments, arguments and criticism. However, a widely accepted explanation is the one that assumes an *appraisal* process as an essential part of the emotional phenomenon, even when the assumptions made on appraisal are different. Moors [175] summarizes the state of the art by arguing that theories of emotion are built on two fundamental assumptions:

> (a) that there are regularities to be discovered between situations and components of emotional episodes, and (b) that the influence of these situations on these components is causally mediated by a mental process called appraisal.
>
> Moors [175]

The former assumption is according with theories such as those referenced here of Northoff [187], López Paniagua [151] or Hayek [105], which consider the relationship between the extern and the inner-Environment of the system. The latter is related with the here subscribed theory of Damasio [54], and assumes that *appraisal* is an inner process that helps to build the *feeling*.[2]

The empirical studies of Northoff centered on to understand the work of the brain's *insula*, suggested a close relationship between the interoceptive awareness and the emotional feeling. Results following this work, provided support for the hypothesis that interoceptive stimulus processing may be involved in the differentiation between independent types of emotional feelings: "Interoceptive stimulus processing is coded in relation to exteroceptive stimuli going beyond mere modulation of the former

[2] At this precise point we are referring *feeling* and not *the knowing of feeling*.

(inter) by the latter (extero)" [187]. By and large, it is assumed some sort of cognitive assessment intrinsic to the emotional phenomenon, even when there are significant differences in the interpretation if the evaluative phenomenon: from those that argue an unlimited number of dimensions and consequently emotional experiences [199], to those appraisal theorists that conceive a set of dimensions of appraisal as a base for the evaluation [71, 221, 244].

Early works of Lazarus [136] argued that emotion arises from the source of continuous processes of evaluation joined to the concept of *value*. Subsequent works of Lazarus show a more concrete analysis about the relationship between the *value* of the situation and the emergence of *emotional reactions* [138]. An analysis about how *emotional reactions* appear under situations assessed as gratings, threats or frustrations, that describes three types of appraisal in terms of a [present] *harm/loss* and [future] *challenge* or *threat*. Lazarus tries to solve the problem of simultaneous types of appraisals related to the same object by assuming different levels of predominance and arguing competition between them [138].

Scherer defends a dynamic model of emotion (that names *The Component Process model of Emotion*) in the attempt to give an explanation about how emotional states arise as a result of a sequence of specified appraisal (i.e. *stimulus assessment*).The *Sequential Check Theory of Emotion Differentiation (SEC)* [237] is formulated as part of this *The Component Process model of Emotion*, in the attempt of Scherer et al. of explaining the differences among emotional states (that are the result of a sequence of defined stimulus evaluation) and the later predictions made on the basis of response's patterns:

> on the observation that the valence, activation, and power dimensions of emotional meaning seemed to be linked to criteria of stimulus evaluation (valence = goal/need conduciveness, activation = urgency, and power = coping potential), I originally suggested a set of criteria (which I called the Stimulus Evaluation Checks, SECa), that are predicted to underlie the assessment of the significance of a stimulus event for an organism (...).
>
> Scherer et al. [237]

One of the key postulates about the nature of these *SECs* is that they are organized "in terms of four appraisal objectives. These objectives concern the major types or classes of information with respect to an object or event that an organism requires in order to prepare an appropriate reaction" [237]. And the preparation for action is related to: (a) how relevant is the object or event (*relevance*), (b) what are the consequences (*implications*), (c) how these consequences are copied (*coping potential*), and (d) what is the significance of the event concerning self-concept and social rules (*normative significance*).

By and large, there is a widely accepted identification of *appraisals* as some type of facet over which the emotional meaning is built. And that it is related with other elements that appear among theories: *value*, *concerns* and *goals*. The computational problem now is how to consider this in order to obtain analogous models in artificial systems.

Thesis Rationale ♣ 11 *Appraisal Assessment.*
We agree appraisal as an essential facet of assessment of artificial emotion. Among the multiple visions of appraisal assessment, we will accept the one of appraisal theorists that conceive a set of dimensions of appraisal as a base for the evaluation [71, 221, 244]

4.4.1 Computational Appraisal

We will assume computational appraisal by referring to the extent we are measuring the effects that some object or event convey to the system, on the basis of some elements identified in the state of the art as *valence* (i.e. value), *appraisal dimension*, *concerns* and *goals*. In order to define a computational theory, to define the way in which these specific terms can be conceptualized is essential. More precisely, the aim of this paragraphs is to describe those theoretical principles that can establish the ground for their computational definition.

The core assumption of Ortony et al. [199] is that emotions represent *valenced reactions* to the objects or events coming from the external environment. And the theory of Ortony et al. value these reactions with qualities such as *pleased, displeased, approving, disapproving, liking, disliking, etc.*. That is, qualitative values that represent the positive or negative relevance that some object is having for the system. As *quality* cannot directly represent usable data in terms of computational systems, we will assume the need of some sort of artifact under which relationships can be built and assessed in order to obtain *quality-like*.

From an engineering perspective, the 'positive and negative relevance' gives the idea of some type of metric or distance, an idea widely accepted in emotion science too. However, rather than solving this metric by means of euclidean distances related to some set of mathematically discretized emotions, we should better find the set and the function that might model the emotional distance. Since metrics are defined to measure distances between pairs of elements within a set, it seems logical to think that, if this metric serves to measure emotion-like distances, it should induce an emotion-like topological property in order to measure distances between emotion-like points. And this relevance is closely connected to the matter of (a) the computational nature of these emotion-like points, (b) the definition of the metric to measure distances, and (c) the definition of those references that allow for feature these distances as 'positive' or 'negative' for the system. And this requirement is directly connected with the concept of *concern*.

Frijda conceived the issue of concerns as a key element for building meaning, and he suggested *the law of concern* as a complement of the Law of Situational Meaning that claims: "Emotions arise in response to events that are important to the individual's goals, motives, or concerns" [81]. Frijda holds that "every emotion hides a concern, that is, more or less enduring disposition to prefer a particular state of the world. A concern is what gives a particular event its emotional meaning. We suffer when ill befalls someone because, and as long as, we love that someone (...) Emotions point to the presence of some concern" [81]. And refers 'world' to either the external environment as well as the own state of the organism, holding that the influence of some emotion is arranged basically by the influence of relevant concerns.

Concern refers thus to some proper function that helps to move the system towards a focus, and helps to convey meaning concerning this focus. Relating this argument to our current work, we will use proper terminology in order to allow the design of an analogous model of the term. We will use the concept of *metafunction* to re-define *concern* from a computational point of view, since *concerns* are related to additional functions of goal compliance.

Thesis Rationale ♣ 12 *Computational Concern (c-Concern).*
We conceive *c-Concern* as an artifact in the form of a meta-function which states the inner reference that uses the system to (a) manage related functions of goals compliance, and (b) measure the positive or negative distance to those matters of interest or importance for itself.

The representational issue that regards the *internal representation* of these references [81] will be solved by assuming principles presumed in the problem of *computational feeling* that will be described later in this chapter.

Finally, we must solve the problem of *goals* and the differentiation from the term *concerns*. Frijda [80] holds that *concerns* produce *goals* and priorities for a system, an idea that gives us the opportunity that makes possible an easy understanding of an emotion-like 'goal'.

Thesis Rationale ♣ 13 *Computational emotion Goal (c-emoGoal).*
The abstract notion of *c-Concern* will produce the notion of *c-emoGoals* related to it, and will set the priorities of a system. Each dimension in which a *c-Concern* is divided. It provides a mean to functionally define and measure the positive or negative error regarding each dimension of a *c-Concern*.

However, this is not enough at all to convey the same biological functionality of concern to artificial systems. The problem now is how to describe the conceptualizations of *c-Valence*, *c-Concern* and *c-emoGoals*, and the relationship of them with the concept of *appraisal dimension*. We have conceptualized the set for measuring emotion-based distances, but distances are not the only computational problem raised herein. We need a description of how the system evolves from one emotional state into another emotional state, and how it is done over the course of time.

We argue that 'time' is an essential element of the emotional assessment nature. It cannot be defined an emotional assessment without the influence of time since assessment itself as well as later processes derived from it are intrinsically affected by this concept of 'time'. Consequently, we argue the requirement of a description about (a) how the system evolves from one emotional state into another emotional state, and (b) how it is done over the course of time.

4.4.2 Computational Process of Appraisal

The problem of appraisal is thus threefold: (a) How to capture the influence of *c-emoGoals* over *c-Concerns*, (b) how the *c-Valence* is built, and (c) how to enable a dynamical system in order to assess stability taking into account the concept of *time*.[3]

Thesis Rationale ♣ 14 *Requirement of an Universe of Emotion.*
We intuit the requirement of an *Universe of Emotion* that integrate *a-Dimensions*, *c-Concerns* and *c-emoGoals*, their interaction and their situational state. This will allow for rating concerning the concept of 'state' rather than the concept of punctual 'value'. There is a problem of *c-Valence* per each of *a-Dimension*, *c-Concern* and *c-emoGoal*.

This section will describe how the process of assessment can be decomposed, in order to build this appraisal space by maintaining at time analogous principles to those that biology holds. To this end, we will accept the idea of Marsella and Gratch regarding the separation of *appraisal* and *inference*:

> In our view, these multi-level theories of appraisal unnecessarily complicate appraisal processes by conflating appraisal and inference. Rather, we argue that appraisal and inference are distinct processes that operate over the same mental representation of a person's relationship to their environment.
>
> Marsella and Gratch [160]

The conceptualization of emotion as a bounded functionality or *'complete background for punctual states together with inferences to driven behaviors'* is not necessarily true. Generally speaking, it seems as if life systems should have linked the emotion triggering and the behavior making (assuming those deviations commonly described in decision-making or behavior essays). However, even when the biological emotion seems to works in this way, the real path is (a) how assessments are made, and (b) how these assessments are monitored and managed to make inferences.

Thesis Rationale ♣ 15 *Assessments and Inferences.*
Under our point of view, the process of assessing is not made within the same functional structure as the process of inferencing (it would bring about unviable implementations regarding computational processes, since the same behavior should be inferred from the source of different values of appraisal). Systems work more successfully if they implement single processes, so it is better to isolate individual assessments (on which to compute individual value-scoring) and let the system uses this value afterwards in subsequent processes such as conducting inference or building complex values.

As inference or behavior are intrinsically related to some essential terms in emotion science such as *emotional tendency*, we will clarify our perspective regarding these essential terms in order to justify our design decision concerning this issue.

---

[3]We are referring herein the influence of *c-emoGoals* over *c-Concerns*. However, this same effect will be accepted regarding the influence of *c-Concerns* over *a-Dimensions*, but it is not described here in order to maintain coherence: the term *a-Dimension* has not still described.

Thesis Rationale ♣ 16 *Emotional Tendency*.
We will assume *emotional tendency* as the normal way to respond with temporal stabilized structures (a.k.a. in emotion science *temperament*). Clearly, *emotional tendency* influences the *emotional process* in biology, and it has been widely studied in order to conceptualize and describe personality traits (see the state of the art). One of the most accepted and used conceptualizations of *emotional tendency* for modeling purposes is *The Big-Five Model—FFM*. It classifies these traits in five dimensions of personality: openness, conscientiousness, agreeableness, extraversion and neuroticism [10]. However, we will not model specifically this feature because we assume that functions, variables, attributes, constraints, etc. as well as several additional factors related to models design, are conditioning in such a way the work of these models that we can assume the 'existence' of this *emotional tendency* in the artificial system. That is, since we are designing an artificial system, we are conditioning from the very moment of design its tendencies; we do not need human-based tendencies in our work. This feature might be added in those cases which might require it, on the same basis of our study.

Focusing now in the very essential process of appraisal, we find a wide and multifaceted state of the art from which we should build our theory. And this theory should involve some explanation for each of these essays in the state of the art.

We accept the vision of Smith and Kirby about the distinction of slow and fast appraisals that work concurrently under different principles. That is, the "two-process model of appraisal distinguishes between slow appraisals based on more or less extensive reasoning from fast appraisals that are associative or memory-based. These slow and fast appraisal processes work in parallel and are integrated to arrive at overall appraisal of an event" Smith and Kirby [242]. However, we do not accept a discrete conceptualization of (a) slow-appraisal and (b) extensive-appraisal.

Thesis Rationale ♣ 17 *Concurrent Appraisals*.
We will accept the work of appraisal under a mixed perspective in which: (a) processes of slow and fast appraisal are concurrently occurring, and (b) these processes are not discrete but they are considered as different appraisals working at the same time. According to the idea of Ellsworth and Scherer [71], we will assume that a system will not assess at the same time two essential *c-Concerns*, and that regarding these concerns appraisals occur sequentially. By contrast, we will assume that—during the assessment regarding a concern—there will be several appraisals that will work in parallel and that will take different computational times depending on the resources they need to do the assessment.

Once established this essential difference, we will address the problem of how and what dimensions should be defined in the nature of *c-Concerns*. Ellsworth and Scherer conceives appraisal as performing an essential role in emotion elicitation and differentiation, with no clear position about which the central dimensions of appraisal are. It is only described *novelty*, *valence*, *goal/needs*, *agency* and *norms/values* as part of those major dimensions postulated by different theorists. Clearly, it establishes a computational problem since these concepts pertain to different realities and implement differences in their nature (i.e. abstraction).

However, this work of Ellsworth and Scherer proposes an explanation about of how theorists make sense regarding appraisal and how it performs. And this essay will be our reference since it has served to our understanding about some features of emotional assessment, and how it might be focused in the aim of a computational explanation.

We highlight the reader again about the computational nature of this dissertation, and our requirement of searching for solutions that can be implemented in artificial systems. As we have previously appraised in earlier chapters, we are not arguing that our solution is correlating the real phenomenon of emotion. This work only describes the principles to build an artificial theory of emotion. From this perspective, we understand the requirement of joining the chain-links from the source of the available knowledge. Thus, we have joined several aspects of appraisal theories to conform an appraisal base for our dissertation. We will then not entirely subscribe a theory as unique solution but we agree with various features of several proposals.

I. Principles of appraisal We will show a summary of principles generally accepted in appraisal theories, in order to construct the ground for designing the *Universe of Emotion* where assessments will be made. The essential aspects to build this space are related to the searching of those pillars by means of which emotion is triggered and categorized.

Thesis Rationale ♣ 18 *Appraisal Identity*.
Regarding the discourse on whether appraisal should be considered a former-event or a component-of-emotion [71, p. 575], we will assume the veracity of both arguments. We will accept that (a) previous assessment are essential in order to trigger emotion, and (b) the requirement of having references to allow for assessments results in dimensions, since these references are establishing principles for each emotional dimension (since these references should be closely related to the emotional concerns, they should be part and consequently a component of the emotional phenomenon).

Thesis Rationale ♣ 19 *Continuous Appraisal and Emotionality*.
We will subscribe the idea of a continuous process of emotion that changes as appraisals are added or revised ([71, p. 575]) with no necessary boundaries between categories of emotion. Even assuming the existence of appraisal combinations (i.e. patterns) and profiles related to the essential categories of emotion, we also agree the idea of a continuous assessment with no need to complete a profile related to some emotion name (i.e. category), allowing in this way for *emotionality* in the sense the work of reference does (that is, the feeling of emotional states with no categorization). This way, we allow systems to 'feel-like' emotion even without a full complement of appraisals is in place, as well as the chance of having *moods* (since they are positive or negative states with *valence* but without defined boundaries). Additionally, as explained in our work of reference, even when it requires individual emotions as bounded categories, it also allows different varieties for the same category (i.e. different types of fear to give an example).

We understand emotions as specific perceptions that comes from the relationship between global values and the system, are intrinsically conditioned by the *time*. This *time* constraints the emotion triggering interval, and it is as long as the time required to recover stability in the global values and the system.

Thesis Rationale ♣ 20 *Time.*
We argue that 'time' is an essential element of the emotional assessment nature. It cannot be defined an emotional assessment without the influence of time since assessment itself as well as later processes derived from it are intrinsically affected by this concept of 'time'. Consequently, we argue the requirement of a description about (a) how the system evolves from one emotional state into another emotional state, and (b) how it is done over the course of time.

We thus subscribe the concept of time as an essential feature of emotion in order to allow the emergence of some appraisal dimensions such as *copying*, as well as re-appraisals that might result in new emotions and that might help control the state of *motivation* of the system.

Consequently, once studied the discussion made in this work, we will assume the following ideas: Regarding appraisal and emotion we will assume: (a) the assessment as the essential core of emotion triggering but not the only cause, (b) the idea of appraisals as components of emotion, since those dimensions over which the assessments are made constitute the essential core of the emotion categorization, (c) the continuous assessment under the idea of a continuous character of emotion, (d) an agreement with the existence of emotionality, that is, feelings that are not filling the pattern of a concrete category of emotion, and (e) time as an essential appraisal's feature.

II. The universe of emotion Under a general agreement, it is accepted that organisms continuously assess stimuli for their significance:

> The central feature of an appraisal perspective on the elicitation and differentiation of emotion is the assumption that organisms constantly evaluate stimuli and events for their significance for the individual. This significance is operationally defined by a number of dimensions or criteria which constitute the meaning structure in which the evaluation takes place.
>
> Ellsworth and Scherer [71]

Taking into account and accepting this argument, we will build our appraisal theory upon a logic of knowledge in which we will represent our hypothesis as a relation of features integrated within the same framework. This framework will be the *Universe of Emotion*, an algebraic-like space where vectorial-like dimensions will assume the role of *appraisal dimensions*, and within which categorized emotions will be defined under the basis of hyper-shapes within this space.

It is highlighted that each type of artificial system will have a significant influence on, and determine the manner of defining appraisal-dimensions. However, we will use the ones defined in our work of reference such as novelty and pleasantness, copying, or social dimensions (among others) for the shake of clarifying our explanation.

Thesis Rationale ♣ 21 *Universe of Emotion (I).*
We conceive an appraisal dimension vector model for representing emotions by means of vectors of appraisal attributes. This way, each design can be structured under those algebraic principles that better fix the requirements of each system. We will name each dimension of this space *appraisal dimension (a-Dimension)*, and each vector of appraisal attributes within this space will be referred as an *Emotion Attribute*. It will be described in the following chapter the construction of this vector *Emotion Attribute*.

Even when common emotions are viewed under similar principles regarding logical structures, this proposal allows for deviations depending on the final generic system *AGSys* that will deploy emotion. Each system can deploy its own emotional principles under the ad-hoc structure that has integrated and that it will use for emotional assessments.

For the sake of a clear explanation, we will describe the model of an appraisal space by using the appraisal dimensions (*a-Dimensions*) and the final emotional assessment as our tools for explanation. However, it should be taken into account that, under our theory, *a-Dimensions* derive *c-Concerns* and *c-Concerns* derive *c-emoGoals*. In such a situation, this space that will be herein described, should be interpreted as one of the instantiations that should be made within the final emotional system *(ESys)* (since there are required as a minimum of two additional spaces where assessing of *c-Concerns* and *c-emoGoals* be made).

Thesis Rationale ♣ 22 *The Universe of Emotion (II).*
We propose a three-dimensional model of assessment to represent the complete problem of appraisal. Our proposal is the building of an universe of assessment, by means of the use of the three main artifacts of appraisal evaluation (i.e. *a-Dimensions*, *c-Concerns* and *c-emoGoals*). This way, the system is able to build an emotional evaluation within the complete universe of the appraisal problem.

This framework allows for partial equal-emotions in quality or strength, classification and rate of emotions according to their possible relevance concerning appraisal dimensions. Now, our final objective in explanation is how the proposed conceptualizations (our proposed line of reasoning the emotional knowledge) are structured and used by systems. The cohesiveness of the structure of conceptualizations, is essential for the operationality of the pattern of interrelations among each computational element herein proposed, and through their linkages.

Thesis Rationale ♣ 23 *c-Valence.*
We refer to *c-Valence* in an analogous sense to that expressed when *'valenced reactions'* are argued in emotion science. It is a different concept from that of *v-State* that refers to a more high level state that integrate relationships (i.e. abstractions). This *c-Valence* is used to capture the value of the positive or negative error concerning some *c-emoGoal*, *c-Concern* or *a-Dimension*. AGSys might perform a valenced reaction related to the consequences of an event, agent or object (by doing reference to the OCC model as an example). These valenced reactions regards to either goals, concerns or dimensions, and they are part of the *v-State*. The proposed *c-Valence* captures the stability degree either be in *c-emoGoals*, *c-Concerns* or *a-Dimensions*.

III. Structure of emotion and emotion categories We propose this framework to allow for computing the relevance of emotion in an *a-Dimension* using the explained hypothesis about the differentiation of emotion categories, and assuming similarities in the pattern of appraisal dimensions for each category. This way, it is also allowed the possibility of (a) having deviations in the appraisal pattern for a concrete category, (b) let the possibility of varieties for the same category, and (c) computing a continuous degree of similarity between categories of emotion.

We will assume that labels naming each category will be part of learning-based processes by which the system will identify the label with the pattern of stimuli.

Usually, animals do not need the knowledge of these labels (i.e. happiness, fearness, anger, etc.) in such a way humans do. However, if we need a human-like system, this is the way in which emotions can be identified by the system.

However, in order to structure the assessment, it is required to define the logical structure under which a-Dimensions, c-Concerns and c-emoGoals are defined. Since this dissertation has been studied under principles of generality, our model allows for the implementation of different structures depending on ad-hoc requirements of each final system.

However, in order to explain our vision about this structure and the way *a-Dimensions*, *c-Concerns* and *c-emoGoals* might be defined, we will use the most common theory in computational emotion. We refer the essay of Ortony et al. that structure emotions under subdivisions based on the consequences of events, action of external agents or aspects regarding external objects, over the system [199]. Ortony et al. establishes a taxonomy, the OCC model, that can be used to situate the emotional state of the system, regarding the valenced reaction to the consequences of the event, the agent or the object perceived.

The still non realizable construction of an intelligent system with analogous intelligent capabilities to those of humans, might allow for small structures that might continue growing from the base of the experience of the system. We then need to build a successful structure for general emotion types, in order to allow for inferences.

It will allow as well for defining concrete *a-Dimensions*, *c-Concerns* and *c-emoGoals*. The emotional structure of emotion depicts a portrait of the emotion types, specializations and generalizations, as well as the representation of consequences (or inferences) under the form of usable laws.

The way each *a-Dimension*, *c-Concern* or *c-emoGoal* state is affecting the system is just an initial data with no meaning for the system. But it is required to be classified under some sort of model that allow the system for meaning regarding the states of these artifacts, and the stage of representation (i.e. the growing towards feeling) in which meaning is built (this is labeled as specialization of emotion type in OCC).

4.4.3 Measurement of Compliance

Any *v-State* will be defined by means of a concrete appraisal state, computed within the *Universe of Emotion*, and the correspondent *c-Valence* (which represents the positive or negative relevance for the system). This relevance, as it has been previously described, is determined by the stability degree of either *c-emoGoal*, *c-Concern* and/or *a-Dimension*.[4]

As we have assumed that life organisms continuously evaluate stimuli for their [individual] significance (as a central feature of appraisal perspective), we will assume the existence of a defined number of appraisal dimensions that will conform

[4]The concept of *v-State* should be associated to the concept of *episodic memory* too, and it will be described later in this dissertation.

the base for this evaluation. Each *a-Dimension* can be assumed as a concrete-matter space where a set of related *c-Concerns* are defined, assuming each *c-Concern* as a base for defining the set of *c-emoGoals*. It is an analogous perspective to that of: *a-Dimension* is a viewpoint[5] of wellness where can be established several views[6] or *c-Concerns* of wellness for that viewpoint and where the wellness for each view is defined by means of *c-emoGoals*.

This way, to measure the consequence of an event under the perspective of a concrete *a-Dimension* such as 'novelty' [71, p. 575], we can define those concerns of the system that might regard 'novelty'. To give an example: one concern might be the derivative of sensor gains, that is, the sensitivity to change of sensor quantities. This concern will be defined by means of *c-emoGoals*, each of one will measure a concrete sensor derivative (i.e. sound, temperature, light, etc.).

I. Influence of time Till now, our theory does not seem to be enough to explain some additional features of an emotional system. Additionally to the direct assessment of an event, there are high-order appraisals (and consequently emotions) closely related to the trajectory of the system over the time. The question of time does not refer to the question of letting the system an interval of time by design to resolve challenges. It is a question that regards the influence of this time on the emotional assessment as an additional variable. Regarding this debate, the following assertions made by Ellsworth and Scherer [71]—while explaining the appraisal dimension of *Coping*—are relevant for our argument:

> Appraisal is proactive, going beyond the immediate situation and assessing the probability of possible outcomes by taking into account the ability to change the situation and its consequences. The ability to cope with a stimulus event can be seen as the ability to free the emotion system from being controlled by this particular event to reestablish a new equilibrium. This does not imply that the organism is necessarily able to reach its original goals; it may modify them, postpone them, or give them up altogether. The major function of the power or coping appraisal is to determine the appropriate response to an event, given the nature of the event and the resources at one's disposal.
>
> "...all appraisal theorists postulate a dimension called agency, responsibility, or causation (...) reflecting the determination of the agent (...) and the cause (...) of the event."
>
> "The attribution of causal agency, whether or not it is accurate, influences the organism's appraisal of its ability to deal with the event and its consequences. This dimension, postulated by all appraisal theorists, is often linked to the general notion of controllability or coping ability."

Whatever the theory assumed to model coping and power, the system describes a trajectory over time: (a) to search for the cause of the event, (b) to look for those resources to deal with it, and (c) to measure the success or failure. The evaluation of whether some given situation is controllable requires evaluations about available resources, re-appraisal of the situation, measurements of compliance and so on.

---

[5]Viewpoint on a system under ISO/IEC 10746-1:1998, 6.2.2 is an abstraction that yields a specification of the whole system related to a particular set of concerns.

[6]View on a system under ISO/IEC 10746-1:1998, 6.2.2 is a form of abstraction achieved using a selected set of of architectural constructs and structuring rules, in order to focus on a particular concern within the system.

In the same way, as *valuable-State* of a system is conceptualized by means of semantics (and this semantics is related with the state of the appraisal dimensions), it is also essential that the conceptualization of *c-emoGoals* be conducted in a way that allows for the evaluation of higher emotional states regarding their own evolution over time. Thus, it is an essential requirement that *Appraisal-Dimension* as well as *c-emoGoals* be modeled by means of some function which outlines the time dependence of an appraisal's state.

> **Thesis Rationale ♣ 24** *Influence of Time.*
> The question of time does not refer to the question of letting the system an interval of time by design to resolve challenges. It is a question that regards the influence of this time on the emotional assessment as an additional variable.

II. Model of goal assessment From our point of view, a computational model of emotion must explain both the rapid dynamics of some emotional reactions as well as the slower evolution of emotional responses that may follow deliberation and inference.

The inner goals of an emotional system are herein identified as views of an *a-Dimension* viewpoint, and conceptualized as *c-Concerns*. These *c-Concerns* are defined by means of a set of *c-emoGoals* that will assess those different parts of the system related to some *c-Concern*. The notion of an emotional intrinsic goal, gives us the idea of movement towards a concerned direction. Concerns will be thus a static description of matters of interest for the system within some appraisal dimension.

> **Thesis Rationale ♣ 25** *Appraisal Stability.*
> By assuming the existence of several concerns under the same appraisal dimension, we can think in some sort of stability concerning the state of each intrinsic goal or emotional concern. If we repeat this argument under our terminology, the *c-Concerns* state influence the state of the *a-Dimension* where they are integrated. But we cannot refer a direct influence of each *c-Concern* over the *a-Dimension* but the stability concerning the relative state of these *c-Concerns*. The set of *c-Concerns* (integrated within a concrete *a-Dimension*) are working towards a concerned direction defined by *a-Dimension*, and what concerns that direction is the stability to maintain the movement towards the direction. And the problem of *c-emoGoals* regarding *c-Concerns* allows for an analogous argument.

Under our proposal, *c-emoGoals* are measuring end-points within the system in order to control errors with defined set-points as a reference. They are continuously measuring errors and changing in value. It should be desirable that these measurements, might allow us for additional knowledge rather than just the value of a concrete error. Emotion science claims for some type of connection between the consequences of an event and memory resources to anticipate knowledge. And it is also essential the requirement of letting the system be aware about that consequences in order to act regarding them. Even when the issue of awareness will be attended later in this chapter, we will refer by now to an essential awareness that regards 'error perception' (since at this level of explanation, the consequences of an event can be measured by means of the classic concept of error).

Our concern now is how the system is able to connect with memory resources from the measurement of an error, and which strategy should allow for let the system be aware about not only the error, but the consequences of this error in its neighborhood. As each *c-emoGoal*, *c-Concern* and *a-Dimension* seem to have three essential roles (the measurement of the error concerning a reference, memory recall and dynamics), it seems as if they might share the same pattern of design. We conceive thus a construct to define these three conceptualizations:

Thesis Rationale ♣ 26 *Four-role model of goals*.
We propose a fourfold-role model of *c-emoGoal*, *c-Concern* and *a-Dimension*, in an attempt to distinguish four different roles of behavior within each of them. The proposed formation is the subdivision of each artifact into four roles in order to achieve: (a) first-order objective: direct measurement of error, (b) second-order objective: relationship between error and memory resources, (c) a third-order objective: an associated dynamical system that evolves over time, and (d) four-order objective: the emergence of a new requirement of recovering error under the form of an emergent objective.

Regarding the associated dynamical system, it is of specific relevance the work of Herrera et al. Concretely what refers to the hypothesis H1: "Emotion emerges from a certain dynamical relationship between embodiment and situation". Our objective is to transform the error in a source of power that can influence the behavior of the system to deal with the error. That is, our objective is to transform this error in movement (without influence of inferences programmed by design) and let the system deals with this movement rather than the error. Freedom degrees in solutions are lower under inferences such as *'IF error THEN do'*, than inferences such as *'IF error THEN power'*, and let the system search for *'do'*.

This conceptualization allows us for conveying to the system some type of data that will push the system to recover the desirable value regarding *c-Concern* or *c-emoGoal*. That is, internally is able to measure compliance. The advantage is that additional processes can be triggered in order to recover the required value. And the reasoning for *a-Dimensions* and *c-Concerns* is analogous.

Thesis Rationale ♣ 27 *Goals stability model*.
The use of differential equations is aimed to provide information previously unknown about the interaction of goals, and to analyze their qualitative effects. The use of this mathematical artifact is proposed in order to provide knowledge about goals behavior and relationships among goals. So that, the stability comes from the equilibrium in the interaction of these goals. And it is an essential informational artifact in order to provide emotional meaning.

Regarding a *c-Concern*, it is proposed to be defined by means of a set of *c-emoGoals*. As *c-emoGoals* has intrinsically associated dynamics, we can conceptualize the *c-Concern* equilibrium as an autonomous system (in a sense of non-dependable of time) [218]. As well, we will allow for studying its dynamics within equilibrium neighborhoods allowing for the integration of relationships among errors within the evaluation of the equilibrium. Additionally, it allows the system for freedom regarding the challenge of raise equilibrium, without constraints made by design. And the reasoning for *a-Dimensions* is analogous.

The use of differential equations is aimed to provide information previously unknown about the interaction of goals, and to analyze their qualitative effects. The use of this mathematical artifact is proposed in order to provide knowledge about goals behavior and relationships among goals. So that, the stability comes from the equilibrium in the interaction of these goals. And it is an essential informational artifact in order to provide emotional meaning.

4.5 Perception

Perception involves a huge set of cognitive processes by means of which the knowledge about an external [perceived] object grows. And it is done by relating this object with essential principles and system laws, during the time that the system interacts with this external object. The theory of perception described by López Paniagua, refers to this as the time during which the context named *environment correlation* keeps on working [151].

It is widely accepted that part of the information that the system integrates by means of the perception comes from emotion. Emotional phenomenon conforms an essential part of the huge set of processes that perception involves. We cannot think of perception without taking into account emotion, nor about emotion without taking into account the phenomenon of perception. Perception is thus an essential issue to attend in this thesis. Our interest in perception is thus focused on getting theoretical tools, in order to build the artifact that can integrate those essential chunks of information that allow for building feeling.

Perception has been broadly studied and there are a wide range of relevant works that we might use in our explanation. However, we will accept the work by López Paniagua within our research group because of its relevance, since it provides those essentials required for our interpretation. Before continuing our explanation, we will firstly conceptualize an essential term for our explanation: the *inner-Object*.

Thesis Rationale ♣ 28 *The inner-Object*.
We will conceive the *inner-Object* as the computational artifact that we situate inside or further in the system, and that represents those descriptive aspects of the [perceived] external object.

López Paniagua conceives "two contexts affecting the process of perceiving", named A and B, that corresponds to (a) the medium where the external object becomes an *inner-Object* concerning the aspect of *environmental correlation (EC)* starting from (A), and (b) the medium where interoceptive and cognitive processes work that concerns the *cognitive equivalence (CE)* starting from B, and that configure the way in which the world is understood by the observer (see Fig. 4.2).

Main differences between the work of López Paniagua and the current dissertation regard the conceptualization of each part of the environment and the system: we talk about *environments* while López Paniagua talks about *context*. Both conceptualizations are intended to separate abstractions, but they do so in different ways.

Our *outer-Environment* is the *distal* environment of López Paniagua (that is, where the external object lives). And our *inner-Environment* conforms the environment where the first abstraction made by the system lives—from the source of the

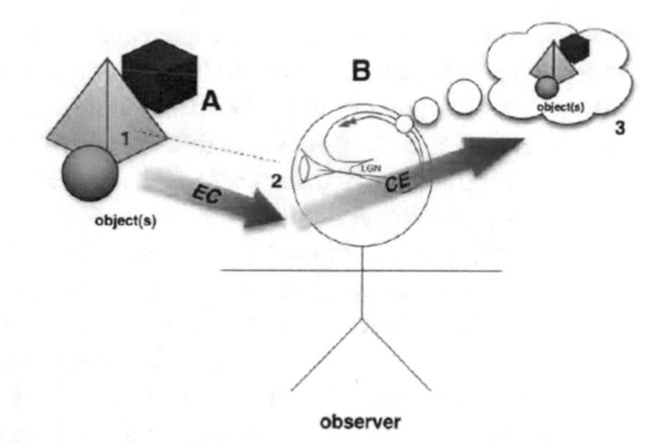

Fig. 4.2 Environmental correlation and cognitive equivalence under the vision of López Paniagua [151]

physical cues and inferences, to computationally recognize a related entity. López Paniagua makes this separation focusing on processes rather than on spaces. Our two environments correlate with spaces A and B in this way: López Paniagua regards A as the *outer-Environment*, and B as the *inner-Environment* of the system.

1. *Environment Correlation (EC)* [151]
   The author describes EC as a correspondence that transforms a distal stimulation into a proximal stimulation to the system, to react and interpret this proximal stimulation afterwards.
2. *Cognitive Equivalence (CE)* [151]
   The author describes CE as a relation of equivalence between the perceived object and the proximal object that comes from the real world. López Paniagua argues differences from system to system in the complexity of CE depending on the system's capabilities and, consequently, differences on the final perception of the same object for different systems.

Under the cognitive approach, we can assume that emotional processes take part in this *cognitive equivalence (CE)*. Emotion, conceived as a result of an emotional assessment, is part of the knowledge that *cognitive equivalence (CE)* conveys to the system. We agree with the argument of Navon [180] that discusses emotions as specific perceptions, which comes from the relationship between global values and the system. As a result, we can hold that perception is the responsible of measuring these relationships regarding references that concern the system.

> **Thesis Rationale ♣ 29** *Perception and Emotion.*
> We hold that perception is the responsible of measuring relationships regarding emotional references that concern a generic system.

Concerning the system, if we refer to an *intelligent system* we are alluding systems capable of deploying intelligent actions. And these intelligent actions are featured with the ability of managing and keeping symbols, that is, abstractions [185]. Mathematically, abstractions are constructions structured from top to bottom to build architectures able to manage complexity, and used from bottom to top to build each new abstraction within the framework of each architecture. Someway, we can argue that perception builds these abstractions by growing information and complexity at each stage of the perceptual process. Whatever the perceptual model might be assumed, we can assume with no discussion that the *inner-Object* is growing in complexity, abstraction and amount of data at each stage of the perceptual process. Consequently, whatever the perceptual model we might assume, the first problem to solve is the architecture of a generic *inner-Object*. This *inner-Object* is required to be transferred from stage to stage through the perceptual process, without problems to be enlarged in order to accommodate its own growing. From the perspective of Bondi, "poor scalability can result in poor system performance, necessitating the reengineering or duplication of systems scalability". This *inner-Object* should increase in complexity in a way that, at each stage, *inner-Object* should integrate the essential information regarding this stage operation.

The same relevance is realized by López Paniagua: the *principle of scalability* stands for "maximizing the degrees of minimal structure, encapsulation,homogeneity and isotropy of knowledge by system growth" in order to achieve property invariance with scale. The term growth is used "in order to remark that the process is a composition of increasing resources and directed resource integration" [151]. Nevertheless, this is an essential feature still unsolved concerning perception representation, modeling and design. There are relevant discussions regarding tools and methodology in order to achieve this property. The own work of López Paniagua that we are using herein, proposes some ideas as starting points in order to solve this issue. Ideas such as the *theory of fractals* [158] in the attempt of capturing the model of growing and integrating abstractions (a) under the same structure, and (b) maintaining the property of compactness.[7])

And we agree that these proposals as essential in order to obtain a computable model of perception.

I. Computable model of perception This conceptualization of *computable model* can be described by considering a brief review of the concepts of computability of Church [47] and Turing [259]. Even when they both refer to *functions* and *numbers* respectively, and their formalizations goes on different phases regarding the ours, they agreed regarding an essential conclusion that affects our problem. This conclusion regards the idea of a systematic method as a requirement for computability. That is, the idea of a general recursiveness or a process of repeated substitution, which is required in order to complete computable procedures.

II. The feature of recursiveness By coming back to the topic at hand, perception, awareness, and several other terms related to human-like intelligence, are argued to be recursive [complex] functions. From the perspective of computer science, this recursiveness is nowadays comprehended as a method by means of which a broad problem solution, depends on results to smaller occurrences of the same problem [92]. And we agree that perception, awareness, emotion, etc. seem to be part of (or even be themselves) some type of recursiveness.

---

[7]We refer here to the property of mathematical compactness to generalize the notion of an entity with information that maintains itself closed and bounded.

Thesis Rationale ♣ 30 *Recursiveness of perception I.*
By assuming perception as a recursive function, we can assume the *inner-Object* as the argument of this function. The debatable question herein is about the conceptualization of recursiveness since, even when the function seems to be characterized by recurrence or repetition, the argument is growing in complexity from repetition to repetition, and this function seems to have different arguments at each stage. Nevertheless, for the sake of an easy explanation, we will conceive this function as a recursive function.

Under this viewpoint, we can argue that perception recursiveness is affected by the growing of abstraction, almost concerning the arguments of the recursive function that might model the process of perception—referring 'argument' herein as the piece of data required by a computational function to operate. And focusing on the argument (i.e. the *inner-Object*), we can argue that it is growing in complexity and abstraction, analogously to the way an integer might grow in quantity under some law of iteration.[8]

We will describe our argument by labeling with *'F'* the function that is systematically repeated under some type of recursion, and *'x'* the argument of this *'F'* that grows in complexity and abstraction by means of some iterative law. Theories such as the one of Hayek maintains a recursive process of relationships integration as the foundation of the perceptual knowledge (i.e. relationships and subsequent consequences). These theories are implicitly arguing the integration of abstractions that are growing during the process. However (as we have previously explained), we cannot assume perception as just a recursive-like function, since the argument of this function is continuously changing in its form. The *inner-Object* is continuously integrating new information built on the basis of new relationships that involve higher abstractions each time.

Thesis Rationale ♣ 31 *Recursiveness of perception II.*
If we thus correlate the idea of *process 'F'* and *perception*, we consequently can correlate the idea of *argument 'x'* with the idea of an *inner-Object*. The idea is that the *inner-Object* used as an argument *'x'* at each perceptual stage, can be used in the following stage in the same shape, but updated concerning information, abstraction and complexity. That is, updated with the integration of new information about relationships, but able to be used by a recursive function that expects arguments in a permanent shape. Thus, *perception* can be conceptualized as a recursive process that sequentially grows the abstraction of its argument.
At each stage this function *'F'* operates with the required level of abstraction, and prepares at time the object *'x'* to be used in the following stage.[a]

[a]This conceptualization is made in order to be usable for our requirements. We highlight that we are not assuming this model as the model under which real perception operates in biology.

We thus agree with this perspective in order to solve the challenge of *cognitive equivalence (CE)* posed by López Paniagua. Our *inner-Object* is equivalent to the conceptualization of that object which derives from the *environmental correlation (EC)*) process. This *inner-Object* is thus the starting object from which *cognitive equivalence (CE)*) begins to work. Perception involves a complex recursive-like mechanism of imagery construction that (a) integrates abstractions, and (b) grows in complexity at each recursion.

[8]Wikipedia [270].

Finally, this *inner-Object* will be conceived as a computational model intended to represent knowledge for the system. This knowledge (to be represented) is argued to be highly abstract in intelligent beings such as occurs in human beings (sometimes these abstractions are referred to as *imagery*). This *inner-Object* thus needs to integrate new abstractions in each stage of the perceptual process. And each of these abstractions will constitute new artifacts to create new abstractions in the subsequent stages, analogously to the recursive process argued by Hayek.

4.6 Autobiographical-Self

Memory is understood in life systems as sets of data that keep essential information for their normal operation. When we talk about artificial systems, this memory constitutes places where data is stored for short or long term retention. Architecture (of both the data and the storage system) is an essential feature that influences its use as an informational system. Memories in life systems are related to experiences and self, and are essential in emotion science in order to maintain references and build feelings [54]. The objective of this section is the comprehension of memory in life systems concerning the influence in building feelings, since it constitutes an essential ground for our objectives.

Conway and Pleydell-Pearce regard *Autobiographical memory* as fundamental in significance for emotion, self and social experience [50]. The study of this feature stands for the same broad complexity that usually share all issues in the field of mind science. Sometimes even similar concepts than those studied in emotion science are studied, such as *personality* or abstractions such as *representation* to give an example. We have chosen the work of Conway and Pleydell-Pearce (a) because of its relevance within its area of knowledge, and (b) because it describes the feature of *autobiographical memory* in the form of a model, in such a way that is susceptible to be used for the purposes of our work.

The main fundamental premise assumed in this work is that "autobiographical memories are transitory dynamic mental constructions generated from an underlying knowledge-base.", an idea that is in accordance with those principles of computer science. Almost all researchers consider an essential relationship between the *biographical memory* and the *self* [50]. However, this work shows additional studies that support this same hypothesis [50]: the *self-referring* feature of the *biographical memory* is what differentiates it from other memories, and some authors argue that they are a 'resource' of the emphself susceptible of being used to make changes of this emphself (i.e., memories seem to be related to personality factors, trait information, patterns of adult attachment, self-schemas, goal change and emotion).

> In short, there appears to be a consensus that autobiographical memory and the self are very closely related, even, according to some theorists, intrinsically related so that autobiographical memory is a part of the self.
>
> Conway and Pleydell-Pearce [50]

4.6.1 The Autobiographical Knowledge

One remarkable characteristic of *biographical memories* under the view of Conway and Pleydell-Pearce, is that they contain knowledge at different layers of specificity. They identify three broad levels of specifity named (a) *lifetime periods*, (b) *general events*, and (c) *event-specific knowledge (ESK)* as shown in Fig. 4.3.

Conway and Pleydell-Pearce explain that "the content of a lifetime period represents thematic knowledge about common features of that period (…) as well as temporal knowledge about the duration of a period", and assumes that each chronological space might be a number of *lifetime periods*. *General events* are showed as more specific and heterogeneous structures of memory, which encompasses both repeated and single events. On the basis of several other authors, Conway and

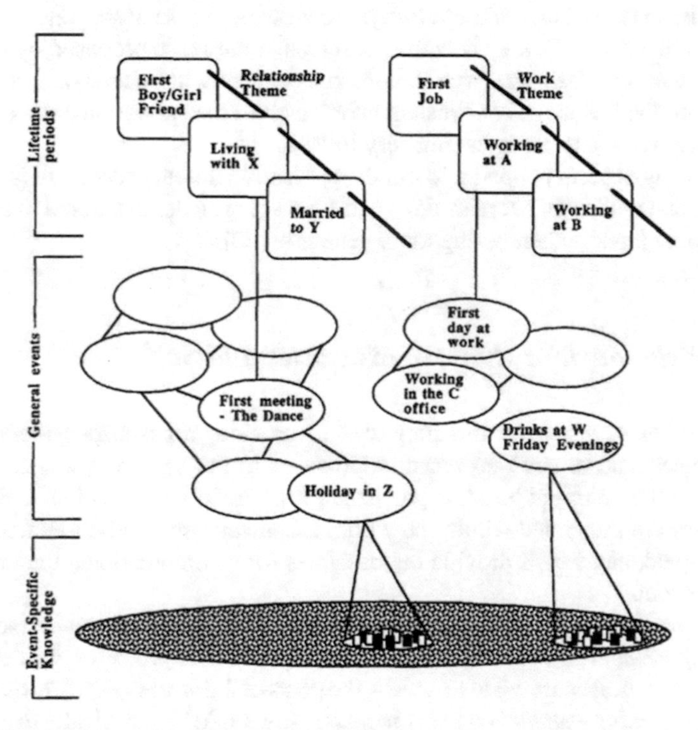

Fig. 4.3 The Autobiographical memory knowledge-base (*Source* Conway and Pleydell-Pearce [50])

Pleydell-Pearce agree that "general events may also represent sets of associated events and so encompass a series of memories linked together by a link.". From a computational perspective, these two ideas are reasoned to be susceptible of being built.

One essential feature of *general events* is that "they feature vivid memories of events relating to the attainment or failure to attain personal goals". In this regards, our thoughts take a look back to the emotion science: the *basolateral (BLA)* region of the amygdala is connected with encoding relevant experiences and has been usually linked to memorable events Mcgaugh [163].

The *event-specific knowledge (ESK)* is a slightly more abstract concept strongly related with imagery. *ESK* is argued to be a predictor of memory specificity and a defining feature of memory vividness. Someway, it conveys the computational idea that we can build *query tools* regarding bases of knowledge or data from the source of the conceptualization of ESK. What represents the challenge is the conceptualization of *imagery*.

However, *ESK* is agreed to be responsible of linking imagery with *general events* (it will not be alluded deeper analysis of the state of the art about weaknesses of these links along time).

The perspective of Damasio about *feeling* and *emotion* is that we [humans] mentally represent any change in our organism state by means of neural patterns and resulting images. And that, if those images occur with a sense of *self* one instant later (and also are enhanced), they become conscious in the sense proposed by Damasio. That is, *feelings of feelings*. Even though this concept will be analyzed some paragraphs later, the feature of "one instant later" justifies the validation of those links to recover *general events* from the imagery in *ESK*.

These images seem popped into mind, apparently without order at all in response to signals that make contact with *ESK*, and *ESK* are even demonstrated to be central to autobiographical remembering more generally [50].

4.6.2 Relationships Between Memories and Self

Autobiographical memories are suggested to provide "a basis for generalizations about the self and others", so recent approaches to the *self* thus suggest ways in which this relation might be conceptualized [50]. The following points and conceptualizations are integrated within the work of Conway and Pleydell-Pearce that we have accepted, and it will provide the baselines for a computational understanding of the problem.

A. The Working Self: Authors use this term to make a connection with the concept of *working memory*, a brain's structure that integrates core processes for controlling separate systems. According to this view, the goals of the working self are assumed as a subset of *working-memory* control processes, structured hierarchically to constraint cognition and behavior. Conway and Pleydell-Pearce conceive the overall goal structure as a mental model of abstract capacities and functions. Authors show additional conceptualizations such as those that propose the existence of *self-schemas* as factors to constraint cognition and behavior, that generate "possible selves" when activated. Focusing on providing an explanation to the nature and emergence of working-self goals, Conway and Pleydell-Pearce hypothesize that working-self goals are essential artifacts to control self-discrepancies established in *autobiographical memory*, being this *autobiographical memory* a knowledge-base that "limits the range and types of goals that a healthy individual can realistically hold".

B. The Working Self Domains: Descriptions about the domains in this work of Conway and Pleydell-Pearce come from the theory of [108]. Three separated domains are described here: (a) the *actual self* as some accurate representation of one's self, (b) the *ideal self* establishing what the self aspires to, and (c) the *ought self* related to what has been learnt throughout life from an educational basis. Conway and Pleydell-Pearce argue that "discrepancies among the three domains lead to characteristic forms of negative emotional experience (...) and self-discrepancies have developmental histories". This argument establishes a good computational framework regarding our objectives, since it establishes the base for building and using dimensions to measure states.

C. Discrepancies regarding Self Domains: Conway and Pleydell-Pearce defend the requirement of functions to reduce discrepancies conceived as negative feedback loops (in the same form as those that are conceived in *Theory of Control* by authors such as Klir [127]), where inputs represent 'the state of the world' that is compared in order to reduce the discrepancy. This comparison is conceived under the evaluation of a complex structure of competing goals (such as those proposed by Oatley et al. [193]) that are responsible for the generation of emotional experiences.

D. Working Self and General Events: Conway and Pleydell-Pearce argue that "vivid memories often arise in response to experiences in which the self and goals where highly integrated (...) or strinkly disjunct". And hold that "the extent to which individuals were able to effectively use appropriate cognitive reactions to deal with dissonant memories was positively related to their sense of well-being, suggesting that control of memory may have far-reaching implications for mental health".

E. Emotions and Memories: Conway and Pleydell-Pearce describe the 'action-guidance system' of Carver and Scheier [44] as a model of the relationship between the abandonment of a goal-attainment and emotion. This system is monitored by a second system to *assess and modulate the rate at which the goal system reduces discrepancy*. The following consideration is essential regarding this thesis: this second system is argued to be *the emotion system, and positive emotions reflect an acceptable rate of discrepancy reduction, whereas negative emotions reflect an increasing failure to reduce discrepancies.*

E. The Communicative Theory of affect: The perspective hold by Conway and Pleydell-Pearce becomes a perspective where goals and plans "communicate with other processes and structures by means of their output, and other parts of the cognitive system communicate with the goal only via its input, as in the negative feedback loop". Even when we presume high complexity concerning this negative feedback loop, the final idea it conveys is the control of goals and associated plans under some complex set-points focusing on the maintenance of their stability. When changes in the probability of achieving an essential goal are detected, the monitoring mechanism transmit alert signals to the broad cognitive system that "sets it in readiness to respond".

E. Theory of affect and James: The argument found in this work that assumes "By this communicative theory of affect it is the alert signals from the monitoring mechanisms that are experienced as emotions" [50], is surprisingly in accordance with the controverted theory of James [118] (subscribed, accepted, discussed and denied), who establishes the origin of emotions at the sense of bodily signals.

4.6.3 Consequences of Relationships Between Memories and Self

Once accepted (a) the work of the emotional system to reduce discrepancies, (b) the influence of discrepancy's control to recover memories, and (c) the fact that those memories are essentially linked within a structure of three layers (i.e. lifetime periods, general events and ESK), it is easy to recognize some essential processes that occur in emotional phenomena.

The usual view in computational and artificial systems is the influence of an event that comes from an environment over the system. Commonly, the system assesses the event's relevance concerning emotion to, afterwards, trigger an emotional process in order to value the object and relate it to the current state of the system and the environment. Essentially, this is not incorrect at all.

> **Thesis Rationale ♣ 32** *Consequences of relationships between Memories and Self*.
> The conceptualization of the autobiographical memory allows us to provide a computational means to implement mental-like processes that might trigger emotional actions such as bodily changes. Since biographical self is joined to the self goals, and the emotional system acts as a control system to maintain discrepancies under control, the whole structure (i.e., goals control under the emotional system and autobiographical memory) is a complete system that might convey signals from part to part. In this way, any thought (i.e. image, abstract representation or so) that can become any estimation of goals discrepancy, will trigger the same processes as if this discrepancy is caused by an event that comes from the real environment. As well, it might be possible the characterization realized by Conway and Pleydell-Pearce: the representation of experiences separately concerning emotional features, to afterwards recall selectively.

4.6.4 Self-Memory System (SMS)

Conway and Pleydell-Pearce hold that memories are featured by the knowledge of goals, as well as for emotional experience. Putting the idea of Conway and Pleydell-Pearce under computational terms, *emotion based memories* can be viewed as computational artifacts whose attributes are *knowledge of goals* and *emotion experience* (i.e., featured by *knowledge of goals* and *emotion experience*). First of all it should be separated the concepts of *emotional goals* and *working-self goals*. It seems as it they are referring different realities. Secondly, from the knowledge of goals to the emotional experience, there should be some method that relate these two types of attributes. Conway and Pleydell-Pearce propose two models as the base to build these methods: (a) the model of Carvel and Scheier [44] that relate goal-attainment abandonment and emotion, and (b) the model of Schemes [233] that relate the functions of maintenance, repair and change goals and emotional experience.

The *Self-Memory System (SMS)* refers "to the conjunction of the working self with the autobiographical knowledge-base, and it is conceived of as a superordinate and emergent system: (a) It is *superordinate* in the sense that its convergent parts, the working self and the knowledge-base, when conjoined allow autobiographical remembering that could not otherwise occur (…), and (b) it is *emergent* in that it is only when the two components interact that they form a system—both can function independently (…)".

As Conway and Pleydell-Pearce describe (and justify), the interaction of these two parts of the SMS can negatively interact with the normal work of the cognitive system. The critical problem is that if intense emotions are re-experienced (as a consequence of any emotional memory that recalls past emotions), the operation of the cognitive system could be thrown into confusion, affecting the current goal structure. This is a direct consequence of the fact that these memories integrate valuable knowledge "of how change was successfully or unsuccessfully negotiated in the past". However, the requirement of a normal work of the cognitive system implies the requirement of an additional work of the working self. That is, the working self should protect against emotional memories to attenuate the re-experience.

As Conway and Pleydell-Pearce propose that one general function of the *Autobiographical memory* is to ground the *Self*: goals are embedded in the SMS with archival connections in the *knowledge-base* and representation in the *working self*. This way, authors propose a model that does not allow contradictions: anyone can maintain a goal that might contradict the *autobiographical memory*. They propose an example to explain this: anyone without the knowledge of 'one's children' can maintain a goal of 'becoming father'. That is, the set of individual's goals is delimited by the autobiographical knowledge to place consistency and plausible constraints on "what goals can be held by the working self".

Finally, in order to fully explain the interaction between the two parts of the SMS (i.e. the *working self* and the *knowledge-base*), Conway and Pleydell-Pearce justify how the *working self* determines what "autobiographical knowledge can be accessed and how that is to be constructed into a memory". This encompasses the whole work of interaction between the two parts of the SMS related to the interests of this thesis.

Thesis Rationale ♣ 33 *Self Memory System.*
Our objective is to conceptualize those artifacts that any *artificial self* would require, based on computational models and the interaction of those models. Thus, our analysis describes those principles to build a representational-based mechanism that might describe the different artifacts participating in the design of an *autobiographical self*, in order to be used by the *Emotional-System (ESys)*. Thus, how additional concepts postulated by authors such as those described in the work of reference of Conway and Pleydell-Pearce (i.e. semantic, episodic, procedural memory and so) is not so much the current focus, but of the future works derived from this thesis. For now, we conceptualize the SMS (under the theory of reference) as a system that "arranges pre-stored knowledge (...) into a form in which it can be experienced as a memory or recognized as part of the personal past" [50]. That is, how SMS represents accessed knowledge into a stable representation compatible with current goals.

Summary: Model of autobiographical memory The following abstract provides a brief summary in order to show a final portrait of the *Model of Autobiographical Memory*:

> The authors describe a model of autobiographical memory in which memories are transitory mental constructions within a self-memory system (SMS). The SMS contains an autobiographical knowledge-base and current goals of the working self. Within the SMS, control processes modulate access to the knowledge-base by successively shaping cues used to activate autobiographical memory knowledge structures and, in this way, form specific memories. The relation of the knowledge-base to active goals is reciprocal, and the knowledge-base "grounds" the goals of the working self. It is shown how this model can be used to draw together a wide range of diverse data from cognitive, social, developmental, personality, clinical, and neuropsychological autobiographical memory research.
>
> Abstract of Conway and Pleydell-Pearce [50]

4.6.5 Grounds of the Self

> Def: autonomy (n.)
> From Greek autonomia "independence," noun of quality from autonomos "independent, living by one's own laws," from auto-"self" (see auto-) + nomos "custom, law"
>
> Harper and MacCormack [101]

Somehow, the *self* concept is intrinsically linked to one's capability of doing something on their own. Usually we refer to the capability of changing rules when we refer to artificial systems, but it is not enough to completely describe this functionality. Additionally, it can be also argued that emotion influences this *self* in several dimensions and that, inversely, emotion cannot be conceptualized without the concept of *self* [54]. Hernández Corbato refers to autonomy in systems as "the quality of a system of behaving independently while pursuing the objectives it was commanded to", assuming open issues on designing control systems to allow systems work alone [106].

Thesis Rationale ♣ 34 *Grounds of the Self I.*
We argue in this thesis that, one of the essential lacks in these control theories, is the avoidance of the *self-objectives* as a critical feature of this independence while pursuing the objectives that should command. *Self-objectives* are closely associated with those emotional dimensions (i.e. *a-Dimensions*) under which appraisal processes work (and that we have conceptualized by means of *c-Concerns* and *c-emoGoals*).

We argue that one key feature of autonomy (and which consequently affects control for autonomy) is the structure and management of the *self-objectives* structure. We hold that control of discrepancies within this structure is a key corner in control for autonomy. The operation of managing *self-objectives* comes to resolving discrepancies when external events appear, and these external events can regard environmental changes or commands to complete tasks.

However, *self-objectives* are not so easy to conceptualize as it might seem. The control system should attend the management of discrepancies in *self-objectives* under the influence of the constraints defined by the *self* and which relationship they maintain with *c-Concerns*, *c-emoGoals* and *a-Dimensions*.

The concept of a *self-objective* belongs naturally (and is joined to) the concept of *self*, and the concept of *self* requires the conceptualization of some additional artifacts such as the *autobiographical self*. Once detailed the work of SMS in previous sections, the question now is how to conceptualize the relationship between this SMS, the memory and the *self* that allow for conceptualizing this entire work in artificial systems:

> The Self-Memory System (SMS) is a conceptual framework that emphasizes the interconnectedness of self and memory. Within this framework memory is viewed as the data base of the self. The self is conceived as a complex set of active goals and associated self-images, collectively referred to as the working self. The relationship between the working self and long-term memory is a reciprocal one in which autobiographical knowledge constrains what the self is, has been, and can be, whereas the working self-modulates access to long-term knowledge. Specific proposals concerning the role of episodic memories and autobiographical knowledge in the SMS, their function in defining the self, the neuroanatomical basis of the system, its development, relation to consciousness, and possible evolutionary history are considered with reference to current and new findings as well as to findings from the study of impaired autobiographical remembering.
>
> Abstract of Conway [49]

In accordance with the SMS model [49], the autobiographical memory knowledge-base in long-term memory contains two types of representations: (a) *autobriographical knowledge*, and (b) *episodic memories*. This SMS is the framework that emphasizes *memories* and *self*. As previously described, *memories* are featured by (a) the knowledge of goals, and (b) the emotion experience. And these two features are related one to each other by methods based on models, in order to assess the emotional reaction concerning (a) the goal-attainment abandonment [44], and (b) the functions of maintenance, repair and change [192].

This SMS is formed by two parts: (a) the *working self* and (b) the knowledge-base that grounds the *working self*. The work of Conway argues that *self* is conceived as a "complex set of active goals and associated self-images" whose work is wrapped within a conceptual sheath named *working self*. The relationship between this *working self* and the *long-term memory* (i.e. the knowledge-base) is argued to be reciprocal. And this relationship is featured by (a) the work of the *autobiographical knowledge* that constraints what "the self is, has been, and can be", and (b) the *working self* that modulates access to the knowledge.

Thesis Rationale ♣ 35 *Grounds of the Self II*.
In order to attend our computational objective, we thus need to make all these conceptualizations less confusing and more clear. In the end (a) *autobiographical knowledge* grounds the *self*, and (b) *memories* (i.e. long-term memory or knowledge-base) are the data base of the *self*. They seem to be equivalent terms but this cannot be far from the truth.

The *self* is conceived as a set of active goals and associated self-images, and the work of these two features in wrapped and conceptualized as *working self*. The *autobiographical knowledge* constraints the *working self* (in order to maintain logic relationships among the goals), and the *working self* modulates access to the *memories* (i.e. long-term memory or knowledge-base) by generating retrieval models used to guide the search process.

The term *memories* refers to long-term memories which are featured by (a) knowledge of goals, and (b) recordings of emotional experience. And they are related through models to assess the emotional reaction to (a) goal-attainment abandonment [44], and (b) functions of maintenance, repair and change [192]. And this provides the grounds for the explanation of high order appraisal dimensions such as *coping*.

It is highlighted that this analysis describes principles to build a representation of computational feeling, by using the requirements concerning *self* postulated by Damasio. Thus, once conceptualized those relationships required to build this representation of feeling, any other additional feature related to memory that might improve our current proposal will be accepted as future works.

Figure 4.4 shows the knowledge structures in autobiographical model under the vision of Conway. Later versions of SMS conceive the *conceptual self* as formed by *themes* in life story, *lifetime* periods and *general events*, and let the *episodic memory* be part of this model. Early versions theoretically referred to this episodic memory as *event-specific knowledge (ESK)*—described in previous sections. However, it is lately conceived as "summary records of sensory-perceptual-conceptual-affective processing derived from working memory (…) and that they form a separate memory system from the conceptual autobiographical knowledge-base".

That is, a separate system whose representations feature the following characteristics: [49]: (a) retain summary records of sensory-perceptual-conceptual-affective processing derived from working memory, (b) retain patterns of activation/inhibition over long periods, (c) they are predominately represented in the form of (visual) images, (d) represent short time slices, determined by changes in goal-processing, (e) represented roughly in their order of occurrence, (f) they are only retained in a durable form if they become linked to conceptual autobiographical knowledge—otherwise they are rapidly forgotten, (g) their main function is to provide a short-term record of progress in current goal processing, (h) they are recollectively experienced when accessed, (i) when included as part of an autobiographical memory construction they provide specificity, (j) neuroanatomically they may be represented in brain regions separated from other (conceptual) autobiographical knowledge networks.

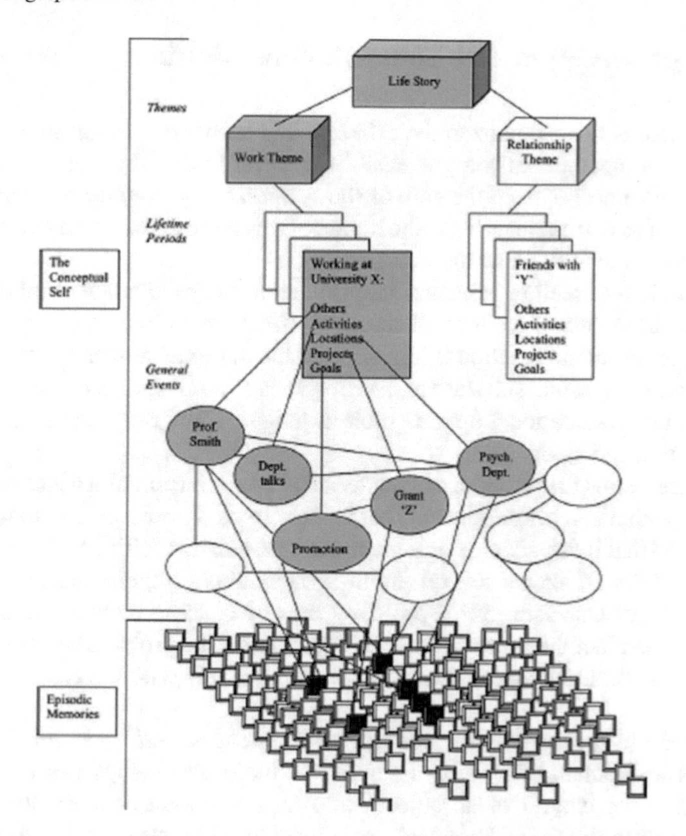

Fig. 4.4 Knowledge structures in autobiographical model (*Source* Conway [49])

Conway [49] refers to the experience of remembering as *autonoetic consciousness* arguing that is a critical component of episodic remembering associated with imagery, sense of the self in the past and type of re-experiencing [50]. And triggering this *autonoetic consciousness* is argued to take place just when *working self* is merged with *knowledge-base*.

Thesis Rationale ♣ 36 *Episodic Memories.*
It is essential the understanding of the role that *Episodic Memories* play concerning emotion. However, the design of analogous computational models is part of a hard problem that defines still open fields of research additionally to the field of enginery.

By now however, it could be modelled uder those artifacts that current artificial systems enable. Regarding the summary records, a single piece of information coming from the context, together with specific patterns of activation can be wrapped under the artifact of an object with partial information. That is, an object that relates partial information from different sources, that is under a concrete control process in order to be stored just when required, and that can be recalled by an event occurring that might regards these sources.

4.7 Representation and Computational Feeling

We have already conceptualized the different key terms in emotion science aiming to allow for computational comprehensibility. Nevertheless, the real interest of artificial emotion emerges from the core of the *symbolic-representation* of *feeling*. The essential problem in engineering is the nature of this representation and how artificial systems might come to know their own emotions.

The problem of feeling is intrinsically joined to the problem of symbolic representation. This is one of the most characteristic features of human beings by means of which, sensorial information is transformed into a higher abstract form that afterwards is used to reason. The state of the art shows several studies made on the basis of different representational options such as language and representation, levels of representation and so on.

This thesis aims the problem of how feelings take the form of a sense, and how it might be symbolically represented. That is, how to convey the feeling to the system reasoning so that it can acquire signification by association with its situational state. The state of the art shows several findings on symbolic representation of feeling. But the work of Damasio [54] is probably the one that can be better used to computationally explain the process, since he firstly treats the problem of consciousness (and how consciousness works) to afterwards relate this with the representation of feeling.

Damasico [54] describes the conceptual problem of *feeling* in relation to two matters: The problem of how brains inside of biological beings cause the mental patterns (i.e. the images of an object), and how the brain concurrently causes a "sense of self in the act of knowing". And he solves this problem by (a) accepting the hypothesis of James [118] regarding bodily changes, and (b) postulating two hierarchical levels of bodily representation on the basis of the mapping of body signals:

1. *The first-order bodily representation*:
   The body is represented in our brain by means of structures that help in its regulation. The external environment is perceived, and the objects are also mapped into our brain by means of the sensory and motor arrangements of perception. These two representations (of body and external objects) are neural patterns that are defined as *first-order maps*. Additionally, he argues a causal-relationship between these two first-order maps (the one related to the body and the other related to the objects) by means of which changes produced on one, induce changes on each other. The representations of these relationships are defined as *second-order maps*
2. *The second-order bodily representation*:
   The problem of how *second-order maps* can become mental patterns or images. The hypothesis holds that, as first and second order maps are related to the body, the mental images that illustrate the relationship are considered to be *feelings*.

Thesis Rationale ♣ 37 *Bodily Representation (on the basis of Damasio).*
We will accept the loop of bodily representation proposed by Damasio [54] in order to computationally describe the conceptual problem of *feeling*. Under identical principles to those of the referred author, we will consider the problem in relation to two matters: The problem of how to build mental patterns (i.e. the images of an object), and the problem of how concurrently causing a "sense[-like] of self in the act of knowing".

We will accept the thesis of James regarding bodily changes. These bodily changes (a.k.a. embodiment) are also argued as essential features within artificial systems if artificial emotion should appear Herrera et al. [107]. Thus, we will take into account bodily-like changes, as well as the two hierarchical levels of representation postulated by Damasio: *The first-order bodily representation* and *The second-order bodily representation*.

I. Feeling The question at issue now is undoubtedly one that concerns consciousness and abstract representation principles. Damasio [54] claims that one cannot experience feeling unless there is consciousness. Hawkins and Le Roux [104] define *symbolic-representation* (a.k.a. as *propositional-representation*) as: "a form of knowledge representation in which arbitrary symbols or structures are used to stand for the things that are represented, and the representations therefore do not resemble the things that they represent" [104].

In the attempt of explaining the problem of feeling, Damasico provided a conceptual analysis of feelings and tried to show how our consciousness works to allow for mental *representation*. Damasio uses the term *representation* making no suggestions about how faithful neural patterns are regarding the objects they refer, and defines the term as a "pattern that is consistently related to something".

He uses this term either as synonym of *mental image* or as a synonym of *neural pattern*, even when he differentiates that *neural pattern* (or *map*) refers to the neural aspect of the process by which a *mental image* is formed:

> ...whatever the fidelity may be, neural patterns and the corresponding mental images are as much creations of the brain as they are products of the external reality that prompts their creation.
>
> Damasio [54]

The images that we can see in our minds "are not facsimiles of the particular [real] object", but a set of correspondences between the physical properties of objects and the dynamic of our organism. This idea is also shared an argued by López Paniagua [151] in this way: "changes resulting from perception have a meaning related to the system activity", and "this meaning is a representation of the state of the universe, relative to the system" [151].

Thesis Rationale ♣ 38 *Maps and Images.*
In order to conceptualize the essential terms of *map* and *image* referred to by Damasio, we will conceive *computational maps (c-Maps)* as patterns of processes that allow for the building of *computational images (c-Images)*. Additionally, will conceive these *c-Images* as patterns of data that can be accessible (i.e. feature of conscious-like) or non accessible (feature of unconscious-like) by the system.

Damasio [54] justifies the importance of consciousness by asserting that it forces you to be aware of emotional bodily changes as a means to construct the feeling. That is, feelings are symbolic representations. One of the essential points in his theory is that feelings cannot be experienced without the work of the biological function of consciousness [54], the operative by means of which we can sense the emotion as happening within our own bodies. And so far, he additionally argues that there is no consciousness that is not self-consciousness, something that terminologically is someway accepted since *consciousness* is commonly used as a synonym of *self-awareness* [106].

> There is no picture of the object being transferred from the object to the retina and from the retina to the brain. There is, rather, a set of correspondences between physical characteristics of the object and modes of reaction of the organism according to which an internally generated image is constructed.
>
> Damasio [54]

II. Consciousness Under the vision of Sanz et al., a system is *aware* if it continuously perceives and generates meaning on the basis of a continuous update of its knowledge (see *Principle 6* in Sanz et al. [230]). Subscribing this, at least three coarse-grain processes by means of which the system perceives, exploits and updates knowledge are required. However, it is not the only issue to attend since the perception of the own bodily changes requires the feature of *self*. The same work by Hernández Corbato [106] maintains that the same system (capable of being aware of something) is characterized as *self-aware* or *conscious* if this knowledge (that perceives, exploits and updates) integrates additional knowledge about its own cognitive system, i.e. *self-knowledge* (see *Principle 8*). Conceptually, these features allow systems for introspection and (consequently) endow systems with the "ability to build symbolic representations of feelings".

The hard problem stated here is the nature of these conceptual features, that is, the nature of consciousness and how systems might build this type of abstractions. Once again, we go back to the theory of Damasio [54] as a reference for explanation. His firsthand judgement of consciousness is consolidated in this hypothesis: "core consciousness occurs when the brain's representation devices generate an imaged, nonverbal account of how the organism's own state is affected by the organism's processing of an object, and when this process enhances the image of the causative object, thus placing it saliently in a spatial and temporal context." [54]. And the analysis is built under the assumption and acceptation of the following essential concepts that we accept:

1. *The somatosensory system*
   It is a "combination of several subsystems, each of which conveys signals to the brain about the state of very different aspects of the body"

2. *The proto-self*
   It is a "coherent collection of neural patterns which map, moment by moment, the state of the physical structure of the organism in its many dimensions" together with the argument that "We are not conscious of the proto-self"
3. *The core-self*
   It is "the sense of self which emerges in core consciousness (...) a transient entity, ceaselessly re-created for each and every object with which the brain interacts" without linking this idea to the one of *identity*
4. *The autobiografical-self*
   It denotes the "organized record of the main aspects of an organism's biography" and it is dependent on systematized memories in which core consciousness has ever been involved.

The proposed hypothesis is that *consciousness* emerges from the base of two essential components:

1. *First component*
   The organism is represented by the *proto-self*. The key aspects of the organism are the person's social environment, the muskuloskeletal frame, viscera and vestibular system. While *first-order neural maps* represent *proto-self* and [external] *objects*, the *second-order neural maps* represent the causal-relationship between these *proto-self* and [external] *objects*.
2. *Second component*
   The first component generates an image that is a nonverbal relationship between objects and organism, and it has two consequences. The first consequence is "the subtle image of knowing, the feeling essence of our sense of self". The other is the "enhancement of the image of the causative object, which dominates core consciousness"

Under the hypotesis of Damasio [54], the *image* of the object created in the *first-order neural map* is joined to the image (causal-relationship) created in the *second-order neural map*. If attention focuses on that object, then emerges the representation of that object in mind. And this is how *core-self* arises.

Finally, Damasio [54] argues that memories come to mind in complex beings such as us by means of complex processes of learning, and it consequently results in the development of the autobiographical memory. It is when these autobiographical records are made explicit in reconstructed images that they become the *autobiografical-self*. This memory is architecturally connected to the *core-self* and the *proto-self* at each time (i.e. lived instant). That is, the nonconscious *proto-self* and the conscious *core-self* are improved by the coexisting display of memorized and invariant facts[9] that come from the conscious *autobiografical-self*. Consciousness conforms the integration of *core-self* and *autobiografical-self*.

[9]Invariant facts are featured by its scopes that can be modified by the experience.

Thesis Rationale ♣ 39 *Essential system elements for Computational-Feeling.*
There are three essential concepts that should maintain its individuality to be distinguished from all other object within the emotional system. That is, three computational concepts that should have entity and should be susceptible of being changed continuously, over the period of time of an object perception. They are the computational analogies of *somatosensory system*, *proto-self* and *core-self*.

These three essential components constitute the base elements to build: (a) the *first component* in order to generate a *c-Image* like a nonverbal relationship between the external object and the physics of the system , and (b) the *second component* in order to generate the consequences of the *c-Image* concerning feeling.

The next question for stepping forward is how to model the processes happening inside the system, since representing the awareness of an object results in different representations besides emotional awareness about that object. Damasio establishes an essential obstacle to solve this problem since we are no conscious of a feeling till we sense that feeling: "…we only know that we feel an emotion when we sense that emotion is sensed as happening in our organism." [54]

Thesis Rationale ♣ 40 *Grounds for computational representation of feeling.*
Damasio poses a question that can be the source of a potential solution: "The process that I am outlining is precisely the same we discussed for an external object, but it is difficult to envision when the object in question is an emotion, because emotion occurs within the organism, rather than outside of it" [54].

It can be a way to envision consciousness concerning an object. We propose thus the conceptualization of emotion under an artifact of the type *inner-Object* and let it grow in abstraction by means of the perceptual process. To this end, we need (a) to conceptualize the perception of feeling as a perceptual process, and (b) an inner-Environment where the first form of the *external-Object* is built under the form of an *inner-Object*. We will name the first effects that this object have over the system as the *emotion-Object*, using an analogy with the name that Damasio employs during his explanation.

4.7.1 The Inner-Environment

We start this section by recalling the thoughts of Hayek: he argued that the environment and the own system are sources of a broad set of stimuli that cause effects within the system. Under his view, the whole set of former effects are the source of some other later processes that are causing new stimuli within the system. From this perspective, stimuli and effects become part of the information to be exploited. This theory supports the proposal made in the previous section regarding the *emotion-Object*: this object might accumulate the information about the consequences that an external object is inducing inside of the system.

Thesis Rationale ♣ 41 *inner-Environment I.*
Theories of perception assume differences in the quality of perception from being to being within the same environmental conditions, at the same moment and for the same type of being [105, 151]. That is, whatever the unique object perceived by two different instances (two different beings) of the same class (type of being) may be, the final image formed in each of the two instances (i.e. beings) is essentially equal in meaning and different in quality. The reason seems to be the different processes that work to build that image (taking into account that these processes have the origin and end within the system). The work of processes in biological beings cannot be thought of as [complete] separated parts, but the nature of artificial systems allows this separation (since they are necessarily designed and built under this conception). Thus, from the perspective of artificial systems, these differences in the quality of perception might be assumed as if they were living within different environments. This way, it might state a first start-point for building a computational conceptualization. In some way, differences in quality perception can be assumed as differences in the references taken to measure features concerning that perception. The contrasting point of view, is that meaning seems to use the same references within the two systems (since they can assume the same significance of the same object regarding the physical world). Quality is formed by distinctive attributes that come from internal references of the systems, and this is why it can be conceptualized as if there were different inner references.

Thus, if references for quality present differences from being to being, we can conceive the artificial systems as if they had two environments. The one that is common to all observers (i.e. the real world) and the one that receives data coming from the inner processes of the system. That is, taking the final perception of systems as the reference for our reasoning, we can argue that two systems that are sharing the same external environment (from an external observer perspective) are [however] living in different [abstract] inner-Environments (from the base of their own perspective). Two systems that are working within an unique external environment realize differences in the final perception of that environment. And these [perceptual] differences can be attended from the perspective of an internal environment.

Somehow the idea of conceptually separating and relating outside from inside processes is commonly accepted, even when the references to this fact are quite different one of each other. To give an example, the *Relational Concept of Emotional Feeling* of Northoff [187] argues that "(…) interoceptive stimulus processing is coding in relation to exteroceptive stimuli going beyond mere modulation of the former (inter) by the latter (extero)".

As this relationship might not be necessarily modulation, arguments like this sustain our hypothesis. An hypothesis that we can assume—from a computational perspective—as an intermediate step between the external object (that causes the emotion) and the emotion itself. Holding this idea, we can argue an individual *inner-Environment* for each punctual system built on the basis of its own references. As we are referring to the measurement of emotion, these references are *c-Concerns*.

Thesis Rationale ♣ 42 *inner-Environment II.*
We propose thus to unfold the system environment into two joined environments: The *outer-Environment* and the *inner-Environment* of the system, allowing the interrelation between them through the conceptualization of relationships related to the influence of the *outer-Environment* over the *c-Concerns*.

From the perspective of design, the *outer-Environment* of the system is the system where it lives (that is, within which it works. The *inner-Environment* is artificially built on the base of previously defined *c-Concerns* and it will be completely decoupled from the *outer-Environment*.

This solution requires the conceptualization of: (a) the *outer-Environment* and its causal connections with the system, (b) the conceptualization of the *inner-Environment* and its causal connections with the system concerns (i.e. *c-Concerns*), and (c) the conceptualization of those relationships that relate both environments.

4.7.2 Grounds for Emotion and Feeling

Once established the computational *inner-Environment* as the artificial environment from which an *emotion-Object* comes from, we still have to face the problem of *feeling the emotion*, since we still "have the obstacle that emotions cannot be known to the subject having them before there is consciousness" (paraphrasing to Damasio [54]).

> To the simple definition as a specifically caused transient change of the organism state corresponds a simple definition for feeling and emotion: It is the representation of that transient change in organism state in terms of neural patterns and ensuing images. When those images are accompanied, one instant later, by a sense of self in the act of knowing, and when they are enhanced, they become conscious. They are, in the true sense, feelings of feelings.
>
> Damasio [54]

The computational viewpoint thus requires a method that enables the system to *'feel'* the *feeling-Object*. Once conceptualized the *inner-Environment* for a generic *AGSys* on the basis of previously defined *c-Concerns*, we can assume that an external object 'exists' inside the system in a different form from the one it has outside, in the real world (i.e. the *outer-Environment*).

I. The problem of representation and the inner object We repeat our conceptualization of a Agent-Generic-System: a *AGSys* is a knowledge-based system capable of deploying sensory-perceptual-conceptual processes, we can argue that our system can represent *abstract objects* from the perspective of Gruber [98]. The concept of abstraction is thus the representation of an *object*.[10] This way, the external object will become an existing object for *AGSys*, in the sense that this object becomes an *inner-Object* that the system can use.

The problem to solve now is the design of those computational processes that, fitted together, might integrate sensory-perceptual-conceptual information about the external object wrapped into an *inner-Object*. And how this informational artifact (i.e. the *inner-Object*) can grow in quantity and abstraction, maintaining at the same time the form, in order to allow perceptual recursiveness.

[10] In the sense of Damasio that comes from the real environment.

Thesis Rationale ♣ 43 *Data type of an inner-Object.*
As this *inner-Object* is required to be used during the whole process of perception (and consequently the process of feeling), it is a requirement that this object can accumulate information usable at each perceptual stage, regarding either the type of data and the functions to be deployed by the process of perception. It will be conceived thus an *inner-Object* as an abstract entity that will properly describe the result of the perceptual process, and that will be able in order to be used by that perceptual process.

Damasio uses the term *object* in a "broad and abstract sense—a person, a place, and a tool are objects, but so are specific pain or an emotion" [54]. And he also distinguishes between *feeling* and *emotion* in two senses: (a) he argues that (likely), from an evolutionary perspective, *emotion* appeared before the *feeling*, and (b) *feeling* performs a long-lasting effect in conscious. And this phenomenon is argued to integrate three stages of processes: (a) "state of emotion", (b) "state of feeling", and (c) "state of feeling made conscious".

The *inner-Object* should cause thus changes within the system. These changes should be represented later by the system, in a way that enables it to perform emotion and feeling, and realizing consciousness of feeling. By referring to representations as *images*, subsequent *images* of this *inner-Object* (that will be created inside the system through the set of interoceptive stages), will be finally related to the self of the system for emotion representation.

Finally, Damasio identifies the problem of inducing emotion, and talks about "ranges of stimuli that constitute inducers for certain classes of emotion" [54]. Without starting philosophical discussions about the convenience of computational emotion, we should define principles and rules about these *computational classes of emotion* that an artificial system might implement, in order to establish the range of stimuli that will induce emotion. We will clarify however our viewpoint regarding the definition of the structure for ranges of stimuli.

Emotion categories has been studied in detail regarding the possible constitution or structure for the purposes of our dissertation. As previously described, we will assume that emotion categories (and consequently, the subsequent labels naming each category) will be identified by the system by means of pattern of stimuli. In order to define those patterns, we propose the choice of commonly accepted theories such as that of Ortony et al. [199].

Thesis Rationale ♣ 44 *The problem of Representation.*
Once analyzed these principles, we can realize some requisites that are absolutely necessary for the objectives of this dissertation. From our computational perspective, we have identified four essential problems to solve in emotion representation: (a) How to induce emotion, (b) what are the essential stimuli in order to induce emotion, (c) how emotion is measured and categorized, and (d) how the systems can feel emotions.

Damasio frames this within four levels of regulation (i.e. basic regulation, emotion, feeling and high reasoning by means of which feeling is felt), and these four frames should be also framed in artificial systems.

The problem can be defined by: (a) how the system perceives to obtain the *inner-Object*, (b) how this *inner-Object* grows in complexity and abstraction to integrate emotional assessment, and (c) how (once the *inner-Object* has grown) is consciously perceived, in order to arise the conscious feeling.

They are all studied in our thesis by accepting those [biological] principles defined in the theory of Damasio. Our aim is to build conceptualizations of these principles that a computational system can be able to use.

II. Inner object for representing emotion and feeling Emotion involves in nature a perceptual process by means of which *'value'* is integrated as a quality, building in this way an essential part of the meaning of an object. Generally speaking, this *perceived value* is essentially related to those *system relevances* that concern the system regarding its body-like state. They are thus references that the system uses to assess the influence of an external object concerning its own state of wellness. And that helps to convey a *perceived value* about that object. To this end, we have previously argued that the system needs knowledge about its body-like state.

We highlight that hereafter in this section, we will not refer to *ESys* or *AGSys*. For the sake of an easy explanation of our viewpoint related to the accepted theories, we will refer to the system now just as a generic system. We will thus use the term *system* as the one able to make emotional assessments and consequently deploy emotion. And we will refer momentarily to these assessments as inner processes. Afterwards in subsequent sections, we will recover our terminology regarding the semantical separation of *Emotional-System (ESys)* and *AGSys*.

As previously described, Damasio uses the term *object* in a "broad and abstract sense—a person, a place, and a tool are objects, but so are specific pain or an emotion" [54]. Under the idea of our proposed *inner-Environment*, we can conceive *feeling* and *emotion* as specific *inner-Objects*.

Northoff [187] argues that an *interoceptive processing* exists that re-represents the interoceptive body state in a unified way by means of which the organism builds a mental image of one's physical state. He holds that this representation constitutes the base for subjective awareness of the emotional feeling, providing in this way support for the hypothesis that interoceptive stimulus processing may be involved in differentiate independent types of emotional feelings.

Thesis Rationale ♣ 45 *The hypothesis of the emotion and feeling objects.*
The statement of *'subjective awareness'* strongly communicates that each system uses its own references (uniquely established within the system) and makes its own representations. Any external object that innerly becomes an *inner-Object* will interact with perceptual stages where lots of interoceptive processes are triggered. Two systems perceiving the same external object will thus infer differences in the qualitative nature of the object. The *'subjective awareness'* will endow different meanings to the *inner-Object* depending on the system where it is conceptualized.

Therefore, as we have advanced lines above, the broad sense in which Damasio uses the term *object*, allows us to conceptualize *emotion* and *feeling* as two types of *inner-Objects*. As Damasio distinguishes between *feeling* and *emotion*, we thus distinguish between *emotion-Object* and *feeling-Object*.

Our argument is that *feeling-Object* is an evolution of the *emotion-Object*, in the sense of abstraction degree. That is, *emotion-Object* and *feeling-Object* will be defined in the form of an *inner-Object*. As this *inner-Object* can grow in abstraction by definition,it will represent firstly emotion (*emotion-Object*) to later represent feeling (*feeling-Object*). Thus, *emotion-Object* and *feeling-Object)* will differ in 'gain' regarding 'abstraction-degree'.

Furthermore,this conception enables subjectivity integration since, by managing the same *emotion-Object*, each system might obtain different *feeling-Object*.

III. Baselines for emotion and feeling Because of the broad laxity of emotion science in terms and meanings, the establishment of correlations with computational terms that will represent an analogous reality in computational emotion is essential. As previously described, when Damasio uses the term *image*, he does so with the meaning of a *mental image* or the equivalent *mental pattern*. And the uses of *neural pattern* or *map* refer to the neural aspects of the process.

Thesis Rationale ♣ 46 *Regarding computational Images and Maps.*
Computer science refers to the use of *patterns* under different viewpoints too, generally alluding to (a) relationships among elements (either data, cues, systems, subsystems, or so), or (b) *states* concerning sets of elements. In order to correlate the conceptualizations of Damasio, we will establish the concepts of *computational-Image (c-Image)* and *computational-Map (c-Maps)*. We will define:

1. *computational-Image (c-Image)*: when alluding to patterns of states concerning a set of system artifacts
2. *computational-Map (c-Maps)*: in order to refer to relationships among a set of system artifacts.

Thesis Rationale ♣ 47 *Conscious versus Unconscious in computer science.*
We will differentiate between those processes that sustain the system, from those that conform the system knowledge. Analogously to biology, we will differentiate conscious and unconscious processes by establishing the difference between (a) system-knowledge and (b) system-sustaining processes. The former are driven to make inferences by the system, and the latter are driven to sustain the operation of the system.

By accepting the view of Damasio, we will assume that some of these constructs conceptualized as *(c-Images)* will be directly accessible by the system in order to be used as knowledge. This is the computational conceptualization of *'being conscious of'* or *'being capable of use'* the information as knowledge.[a] Consequently, we will assume—in reverse way—that some of these constructs conceptualized as *(c-Images)* will not be directly accessible by the system. This is the computational conceptualization of *'not being conscious of'* or *'not being capable of use'* the information as knowledge.

When Damasio refers to 'accessibility', he argues that "conscious images can be accessed only in a first-person perspective (my images, your images)", while "neural patterns, on the other hand, can be accessed only in a third-person perspective". Analogously, we will assume that accessible *c-Images* or *conscious c-Images* will be retrieved only by the system and for the use of the system, in terms of *'used as knowledge-base in order to make inferences'*.[b]

Even when *c-Maps* and *c-Images* are conceived as different artifacts (relationships and states respectively), we argue that they both are used [by the system] in a similar way. They both are used [by the system] to provide mechanisms in the operation of assessment. The inner processes of evaluation use *c-Maps* and *c-Images* to infer results that are used afterwards by the system. Under the idea of Damasio about the lack consciousness of these data, the computational system analogously will not use emphc-Maps or *unconscious c-Images* in order to make inferences.

[a] We highlight hard differences of this conceptualization and the one that refers to the *self-consciousness*.

[b]We are referring herein to inferences that involve methods at the intersection of machine learning, artificial intelligence, bayesian inference, probability, statistics or knowledge-base systems among others.

III. Processes for emotion and feeling Now, the following problem is the establishment of those interoceptive processes responsible of that subjectivity which refers Northoff.[11] The initial *emotion-Object* cannot endow the system the whole subjectivity of emotion, because it does not have integrated essential interoceptive relationships of the emotional assessment. This *emotion-Object* needs interaction with uniquely defined [system] references.

By retrieving the description of the perceptual process, this interaction will be conceived under a systematic work that recursively repeats specific processes. And these specific processes will be featured by the abstraction degree of the stage where they work. The focus thus is which type of [iterative] processes should the system implement, in order to allow for what Damasio names *imagery*. We now refer to *imagery* because of the author argument that "there is no feeling without consciousness", and *Imagery* involves the abstract idea related to *thought* and *consciousness*.

Thesis Rationale ♣ 48 *From inner objects to feelings.*
We will thus assume that an object coming from the real environment is engaged by means of an *inner-Object* within the *AGSys* representational system. Under an analogous idea as such of Damasio, we will assume the construction of this *inner-Object* from the basis of the external object. And we will assume the re-construction of this *inner-Object* in order to integrate higher abstractions at each perceptual stage. The problem now is how this *inner-Object* is computed within our system to create the *emotion-Object* and afterwards the *feeling-Object*.

It will be assumed the conceptualization of a *feeling-Object* in an analogous form as Damasio conceives the *feeling* before *'knowing feeling'*. That is, this *feeling-Object* is conceived as a construct that will constitute knowledge for the system. But this knowledge does not serve so that the system can represent it to itself. It is just when the system is aware of this *feeling-Object* that it can "know that it is feeling". And this concerns the already explained *autobiographical self*.

We are facing thus three problems at this very moment. They are the conceptualization of those cognitive processes through which emotional assessment is made and integrated in order to (a) build the *emotion-Object*, (b) build the *feeling-Object*, and (c) establish how this feeling-Object is related to the *autobiographical self* in order to build the *emotional experience*.

According to Damasio, an object that lives within the external environment (i.e. *(outer-Environment)*) is recognized, and the object is mapped into the system by means of a sensory-motor arrangement. It thus becomes an *inner-Object* living in our conceptualized *inner-Environment*. But from here on, there is a complex process of operation that is required in order to conceptualize the computational feeling.

Since the problem of feeling is intrinsically joined to the problem of symbolic representation, we need the conceptualization of that process that might transform the *inner-Object* in the source of a high abstract artifact that might convey *perceived*

[11]It is highlighted that we now are referring *'inner operation'* to allude sets of processes (either part or the complete set) required for the accomplishment during the *CE* of López Paniagua [151].

value to the system. We need a process for high abstract reasoning from the source of this *inner-Object*.

In order to emphasize the principles for our conclusions, we will remember the essay of Damasio [54] regarding the conceptual problem of *feeling*. He describes this problem of *feeling* in relation to two matters: (a) how the system creates *c-Images* of an object, and (b) how the system concurrently might cause the "sense of self in the act of knowing". And he solves the problem by means of three assumptions: (1) accepting bodily changes as essentials under James [118], (2) postulating a *first-order bodily representation*, (3) postulating a *second-order bodily representation*.

The *first-order bodily representation* is conceived as a complex causal relationship between *first-order maps*, that become *second-order maps* as a representation of these causal relationships. The *first-order maps* are neural patterns that represent the *'body'* and *'the external object'*. The *second-order maps* will represent the causal relationship between the *body first-order map* and the *object first-order map*.

The *second-order bodily representation* is the problem of transforming *second-order maps* in mental patterns or images. The hypothesis of Damasio is that, as *first-order maps* and *second-order map* are related to the body, and the mental images that illustrate this relationship can be considered *feelings* [54].

The computational problem is how to build *computational representation* as an analogy of *mental representation*. That is, under the same principles of biology, the system needs to construct representations that are not reproductions of the real object. Representations will be a set of correspondences between those physical features of the object recognized by the system, and the own dynamics of the system. Once at this point, the following problem is the one that concerns the principles of abstract representation, since Damasio claims that there is not experience of *feeling* unless there is no consciousness (arguing additionally that there is no consciousness that is not self-consciousness).

Damasio judgement is that "core consciousness occurs when the brain's representation devices generate an imaged, nonverbal account of how the organism's own state is affected by the organism's processing of an object, and when this process enhances the image of the causative object, thus placing it saliently in a spatial and temporal context". He conceptualizes four artifacts (i.e. the already described *somatosensory system, proto-self, core-self and autobiographical-self*), in order to argue that *consciousness* born from the base of two essential components: (1) the *First-Component of Consciousness* in which the system is represented by the *proto-self*, and (2) the *Second-Component of Consciousness* which involves the subsequent consequences of the image created by the *First-Component of Consciousness*.

By focusing attention on an object, it is emerging the representation of that object and, subsequently, it is arising the *core-self* (conscious but still without reference to *identity*). It is when autobiographical records are made explicit in reconstructed images that they become *autobiographical-self*, and it is this memory the one architecturally connected to the *core-self* and the *proto-self* at each time (i.e. a lived instant). And *consciousness* conforms the integration of the *core-self* and *autobiographical-self*.

Thesis Rationale ♣ 49 *Grounds for core-self*.
We need to conceptualize the *core-self* as an artifact that will engage the focused *attention on an object* required in this process. We argue that this attention emerges from the act of assessing something that is relevant. As we are focused on emotional relevances, this can be assumed as the act of assessing something that is emotionally relevant for the system. And it comes from the basis of a logical structure of relevances that system deploys and uses to assess.

These logical structures of emotions, viewed as taxonomies of events and emotional consequences, allow for connecting the *proto-self* together with *somatosensory system* and the *core-self*. Under the idea of Damasio "there is at least one other structure which re-represent both proto-self and object in their temporal relationship and can thus represent what is actually happening to the organism". The constraint *'temporal'* assumes an interval of time that exists while changes exist within the system because of the represented object: "proto-self at the inaugural instant; object coming into sensory representation; changing of inaugural proto-self into proto-self modified by object". In order to make an explanation of this relationship, we will retrieve some essential ideas proposed in this chapter.

We have assumed *computational appraisal* by referring to the extent we are measuring the effects that some object conveys to the system. We have assumed that *c-Concerns* are inner references under which the system measure this positive or negative relevance under the theories of Frijda [80]. And that *c-emoGoals* are those values that regard some *c-Concern*, and that uses the system to measure the positive or negative error in order to assess the state of the referred *c-Concern* [81]. We have proposed an *Universe of Emotion* as a collection of *appraisal dimensions (a-Dimensions)* which may be scaled by numbers and added together, and that conforms the number of independent directions of assessment for the system (assuming similarities in the pattern of appraisal dimensions when defining emotion categories). The *c-Concerns* are defined by means of a set of *c-emoGoals* that will assess those different parts of the system related to the wellness of the *c-Concern*.

Each *appraisal-dimension* can be assumed thus as a concrete 'matter-space' where a set of related *c-Concerns* are defined, assuming each *c-Concern* as a base for defining the set of *c-emoGoals*. And the model of a *c-emoGoal* is required in such a way, that can explain both the dynamics of an emotional reaction, as well as the slower evolution of the emotional responses that may follow deliberation and inference. And we have proposed a threefold model for the three constructs *a-Dimension*, *c-Concern* and *c-emoGoal* in which the 'third-order objective' regards an associated dynamical system.

Thesis Rationale ♣ 50 *Computational proto-self*.
The *Universe of Emotion* can be analyzed thus by means of the state of the *appraisal dimensions (a-Dimensions)*. These *appraisal dimensions* can be evaluated by means of the analysis of *c-Concerns*, and each *c-Concern* by means of the analysis of *c-emoGoals*. We argue the requirement of a dynamics analysis, in order to conceptualize the unconscious *proto-self*. The proposed model allows for triggering alarm-calls by means of the analysis of the state variables related to each *c-emoGoal*: x_1 and x_2. This enables the system to trigger basic regulation related to the *c-emoGoals*.

Thesis Rationale ♣ 51 *Computational arousal (c-Arousal).*
Arousal is the state of exciting by means of which some essential processes are triggered in order to optimally face a punctual situation, and constitutes the ground for emotion [5, 54, 107, 118]. We will conceive these arousal states in artificial systems as interruption states, that is, a state in which alerts aware the system about the requirement of immediate attention regarding some source that is establishing high-priority.

In computer science, *interruptions* are signals that are triggered within the systems in order to indicate the requirement of immediate attention. This way, processors are alerted to a high priority state in which they stop processes that are defined as non critical, to attend those that regards as more important under the priority that defines the interruption.

Thesis Rationale ♣ 52 *Computational somatosensory system.*
Interrumption states allow us for conceptualizing some sort of *somatosensory system* based on the state of interruptions of the system that can be used to create the *core–self*. This *core–self* will be conceptualized as an artifact that is argued to focus a relevant event by means of the analysis of the *somatosensory system* state (i.e. the state that system interruptions conform).

Finally—as we have accepted and previously analyzed—the definition of *c-emoGoals* that will end in specific categories of emotion, will be part of models such as those proposed by Ortony et al. [199] (or the subsequent logical form of this theory studied by [248]). Taxonomies such as these, conform the logical account for emotion triggering in origin (without assuming the requirement of the final triggering of this emotion). It is the origin and not the final triggering what now we argue as foundation for conceptualizing the relationship between *proto-self, somatosensory system* and *core-self*.

VI. Bring about feelings We agree with the argument of Navon [180] that emotions can be regarded as specific perceptions that come from the relationship between global values and the system. And we will accept the explanation about the "feelings of feelings", as well as the idea of the body as the shared essence of emotion, feeling and consciousness [54, 107]. But regarding how to bring about feelings, there is required some additional arguments to those described concerning consciousness.

Damasio refers to the conceptualization of the "substrate for feelings of emotion" [54], holding that this substrate arises in two types of biological changes:

1. *Changes related to body state*
   It can be attained by means of two mechanisms: (a) the one that integrates *humoral signals* and *neural signals* and that he names the *"body loop"*, and (b) the one that forwards the representation of the body-related changes before they are really represented, the one named *"as if body loop"*. This later mechanism bypassing the body is the one that [he argues] produces the optimization of behaviors concerning time and energy (allowing for optimal responses) and letting processes of *internal stimulation*.[12]

[12]Damasio references [109] regarding the concept of internal stimulation.

2. *Changes related to cognitive state*
   It inducts specific behaviors, changes in the continuous processing of body states and modifications in the way of cognitive processing.

Thesis Rationale ♣ 53 *Computational substrate for feeling emotion.*
Analogous to the theory of Damasio, we will construct this substrate on the base of two types of computational changes:

1. *Changes related to physical state*
   It will be conceived by means of two mechanisms: (a) the creation of a "physical loop" to assess the cause and effect between the system and those models that regards humoral-like signals and neural-like signals, and (b) the creation of a construct that helps to estimate that under specific representations of feeling, it will regard some consequences in the future. This construct will be conceived as an "anticipative loop" that will influence optimized behaviors.
2. *Changes related to cognitive state*
   It inducts specific behaviors, changes in the continuous processing of physical states and modulation in the processes of assessment.

Somehow, they are dependencies that continue till the result of these changes (related to either body or cognitive states) translates into consequences for the system.

4.8 Architectural Theoretical Foundations

Till now, it has beed described the analysis made in order to support a final model of emotion. It has been proposed a structure on the inside of a model-based cognitive agent *AGys* that will be the responsible of performing the emotional operation. This internal structure is referred as *ESys* and performs the emotional operation on the basis of transversal data concerning *AGSys*.

It will be assumed that reasoning is a matter of *AGSys*, and that *ESys* builds new information about the transversal state which is sent to *AGSys*. And that *AGSys* is the one that interacts with the external environment, while the work of *ESys* is performed within the *inner-Environment* (the one conceived to transversally monitor the *AGSys* performance).

It is thus required to establish the basis for developing a logical model of interaction between *AGSys* and *ESys*, since their reciprocal action and influence is an essential need for the particular purpose of emotion performance.

Since *AGys* is a model-based cognitive agent, it has been assumed a cognitive approach of emotion appraisal, under which it is accepted that which emotions are constructed on the basis of emotion-based assessments. By analyzing the term "assessment", it seems to be related to the idea of reasoning. When it is referred an emotional assessment, it directly implies the existence of emotional references. Once analyzed the assessment itself, the question now is how this assessments is structured within the environment of interaction *AGSys-ESys*.

I. Reasoning capability and emotion Russell and Norvig holds that intelligent agents may implement several components to endow intelligent capabilities (interaction with the environment through sensors and actuators, components for keeping track of the state of the world, projecting, evaluating and selecting future progression of actions, learning, and some several others), while basic agents might not need this same range of features:

> Whereas an amoeba will reflexively move if the temperature changes, a human facing bitter cold will feel distress, experience physiological changes and a strong urge to move, yet may stand fast." "(...) the primary difficulty is coping with courses of action – such as having a conversation or a cup of tea-that consist eventually of thousands or millions of primitive steps for a real agent. It is only by imposing hierarchical structure on behavior that we humans cope at all.
>
> Russell and Norvig [223]

It seems reasonable that—depending on the capabilities of each system—this reasoning will be more or less complex. And these differences concerning reasoning capabilities, can be also assumed regarding emotion—since deployed emotion will be also more or less complex.

Thesis Rationale ♣ 54 *Complex reasoning and emotion.*
We accept that emotional capability requires ability for projecting, assessing and selecting courses of action. Consequently, as far as a system is able to make complex reasonings, the emotions will result in a more complex realizations depending on how featured is the system regarding intelligence.

Focusing on complexity, we previously argued that complex functions (such as those involved in perception and emotion among others), cause difficulties regarding abstractions. And abstractions imply no determination, and decomposition of complex functions into single functions [151]. To manage complexity, the concept of abstraction is used in computer science, and it draws levels of complexity to computationally attend each level separately. But this drawing is accepted not only within the perspective of computer science.

Regarding complexity management, Leventhal and Scherer [147] view the emotional process under a hierarchical structure organized in three different levels of work: *sensorimotor*, *schematic* and *conceptual*. Later works of Scherer et al. followed this same perspective and described the *multilevel sequential checking* [237]. This theory holds that the system posits three levels of appraisal processing (innate or 'sensory-motor', learnt or 'schema-based', and deliberative or 'conceptual'), arguing that (a) the *relevance check* precedes *implication-check*, (b) the *implication check* precedes the *coping potential check*, and (c) the *coping potential check* precedes the check for *normative* significance.

This viewpoint does not spoil what Scherer et al. justifies regarding the work made by each part: (a) the *relevance check* includes assessment of *novelty*[13] as well as relevance to one's goals, (b) the *implication check* includes assessments of 'cause', 'goal conduciveness' and 'urgency', (c) 'coping potential' includes assessments of

[13] This is one of the commonly accepted appraisal dimensions.

control (whether the situation is controllable) and power (whether the individual has the power to control it), and (d) *normative* significance includes assessments of compatibility with internal and external standards.

Thesis Rationale ♣ 55 *ESys based reasoning.*
By accepting this perspective that regards three layers, it will be assumed the work of *ESys* within three levels of work: (a) the work related to the emotional objectives (i.e. *c-emoGoals*, *c-Concerns*, and *a-Dimensions*), (b) the work related to monitor the situational state of these emotional objectives (i.e. the mutual influence among them), and (c) the emotional state regarding the monitored situational state. This way, the emotional assessment (*ESys*) and the reasoning about the emotional assessment (*AGSys*) are improved.

This way, any agent *AGSys* can be assumed a potential emotional agent. That is, it is just required the essential requirements concerning rationale in order to perform emotion. And the level emotion deployed will be related to the capabilities of *AGSys* reasoning rather than the capabilities of *ESys* reasoning. For the same *ESys* result, it will be a matter of *AGSys* performance that the final behavior will be more or less complex.

II. Mutual interaction Hayek describes emotions as *affective qualities* whose order of *mental qualities* does not refer to particular points in the space, as in the case of sensory qualities [105]. He argues that, even when the principles of operation might be the same for the two qualities, the affective qualities represent a temporary preference for certain types of responses towards external situations (i.e. behaviors or a set of following actions), while the sensory qualities represent a spatial relationship for sensory stimuli. Under his viewpoint, there are two different qualities that, however, are working joined.

Thesis Rationale ♣ 56 *Principles of reciprocity.*
We accept the common argument that understands *cognitive processes* as reciprocal items, usually characterized by multiple feedback cycles [77]. And we also agree that emotions provide with quantifiable meaning the cognitive processes. Following the idea of *cognition* as the process of acquiring knowledge and meaning, the *emotion* constitutes an essential feedback which inclines the system to measure and maintain its well-being equilibrium. So that it is essential to think well of these feedback cycles in order to integrate a strategy that avoids inaccuracy in design by erasing essential features of emotional feedback.

With this objective, Fox does a relevant theoretical overview, from which we extract some essential ideas and authors [77]. This overview analyzes the work of Power et al. [211] arguing that they conceive emotion from the same perspective of Oatley and Johnson-Laird [192]: "(…) the function of the basic emotions is to set the cognitive system within a particular framework". If any goal becomes unstable, an emotion suspends the normal work of the system to readjust the values towards a new equilibrium. So that Fox argues (under these theories of Power et al. and Oatley and Johnson-Laird) that "a primary function of basic emotions is to reconfigure the cognitive system so that a new goal can be achieved".

It sounds thus reasonable to accept this related idea: the emotion reconfigures the cognitive system in order to return the system to the equilibrium. That is, there is an action from the *ESys* to *AGSys* in order to obtain a method for *reasoning-concerning-value*.

Thesis Rationale ♣ 57 *Emotion reconfigures Cognition.*
We agree that emotion reconfigures the cognitive system in order to return the system to the equilibrium. That is, there is an action from the *ESys* to *AGSys* in order to obtain a method for *reasoning-concerning-value*.

Ellsworth and Scherer share a similar perspective regarding some other appraisal theorists, about the existence of an appraisal dimension aimed to endow determination, causation or responsibility to the system: *causal-agency* [71]. Even when there is not argued an accurate description about this *causal-agency*, it is however accepted as an influence for the appraisal processes, and linked to the general notion of controllability or coping ability. However, this operation will be assumed in this thesis under the work of those references related to the autobiographical knowledge and the working-self goals. Since they affect to the processes of cognition and behavior, there will be assumed the capabilities of agency and copying as part of these processes. However, it is strongly assumed that its is a simplification made for the purposes of this thesis, and that they are part of the future works in emotion and artificial systems.

Thesis Rationale ♣ 58 *Related to high capabilities and agency.*
Causal-agency will be accepted as an influence for the appraisal processes, and linked to the general notion of controllability or coping ability. This operation will be assumed under the work of those references related to the autobiographical knowledge and the *Working-Self goals*. Since they affect to the processes of cognition and behavior, there will be assumed the capabilities of agency and copying as part of these processes.

Power et al. take into account the key question of emotional disorders, and we will use this theory as well as the overview made by Fox to explain our own perspective. *ESys* as well as the *AGSys* need to be both under control to ensure stability. Basic emotions are linked to emotional disorders [211]. An emotion can turn into exaggerated values that may result in an erroneous work [77]: basic emotion of *fear* might become disorders of *panic* or *specific phobias*, and *sadness* may result in *pathological grief* or *depression* [211].

Fox describes how "we all have been in situations where we have experienced a disconnection between our gut feeling and a more rationale analysis of the situation" [77]. And afterwards, during a discussion about *dynamic systems approach* and the linearity between causes and effects in emotion, she argues that *neuroscience tends to emphasize the complexity of neural systems with a focus on how many different neural structures and circuits can influence each other in bi-directional ways*. This consideration moves this analysis to the argument that dynamic systems "are characterized by wholes out of interacting parts" and refers to the theory of Haken that argues that this occurs by processes related to " 'self-organization' and 'circular causality' ". That is, a spontaneous emergence of order among the elements of a complex system (i.e. *self-organization*) and a bi-directional causality between the layers of the system (i.e. *circular causality*).

Once more, this question moves Fox to a new interrogation about the relationship or even mutual influence between cognition and affect. Simon argues this mutual influence, and holds a system of cognitive-emotional interruptions on the basis of analysis of relevance (even when this analysis was made regarding the work of the attentional processes).

Fox holds that "as different research traditions conceptualize the direction of relationship as occurring in opposite directions" (i.e. cognition-to-affect as well as affect-to-cognition), it seems reasonable that more likely "the relations between these two constructs are actually bi-directional, with causal influences running in both directions". This is a relevant idea that might be justified on the basis of empirical observation: life systems have some sort of aptitude to push an *emotional impulse* back with different levels of modulation, while they maintain their principles of work (usually called *drives* in emotion science).

> **Thesis Rationale ♣ 59** *Mutual influence AGSys-ESys.*
> It will be assumed a mutual influence between *AGSys* and *ESys* in the sense that: *ESys* is monitoring the transversal state of *AGSys* and influences the control of *AGSys* on the basis of its work. And *AGSys* should have some negative feedback in order to reduce the influence of *ESys* under required circumstances.

III. Generalization Taylor et al. discusses about "what concepts should be chosen at the outset of a design task", and solves this by arguing that "search for a simple machine that serves as an abstraction of a potential system that will perform the required task" [257]. The conceptualization to what *single machine* refers, is an essential simple program that entirely solves the problem within some domain: "How does an avionics system work? 'You just read all sensors, calculate the control laws, write values out to the displays and actuators, and then do it all over again".

By considering the work of *AGSys-ESys* as a *single machine*, it can be generalized this approach on any type of system, whatever the architecture it implements.

> **Thesis Rationale ♣ 60** *Single Machine of AGSys-ESys interaction.*
> We propose to assume *AGSys-ESys* as a *single machine* in order to allow for implementation regardless the system. This single machine will be conceived as a conceptual unit of longitudinal-transversal operation on systems, or even on specific part of systems. The operation of *AGSys* can be assumed as the intelligent monitoring of the system (or an specific part of a system), and the operation of *ESys* as the monitoring of wellness for that system.

To give an example, it might be the case of a layered architecture which grows in complexity in such a way as to improve the ability for reasoning. These types of architecture use to implement the physical layer at the bottom layer, and increase in complex reasoning—from scheme to deliberative—while moves up to the top layers. This thesis argues for single solutions instead of large solutions regarding system problem domains. Consequently, under the circumstances of a layered architecture such as the one herein described, it is conceived the attention of each level as an individual problem domain.

Under these circumstances, the use of the conceptual unit *AGSys-ESys* can be described under the repetition of this unit as many times as layers are deployed by the system. That is, an architectural design that displays a repeating transversal pattern between each layer, and the layer above it.

IV. Towards social emotion The conceptual unit *AGSys-ESys* does not only allow for the principle of general applicability. It also addresses the problem of social complex emotions by establishing a starting point of design.

Since systems that implement less layers will deploy simpler emotions than those deploying larger number of layers, systems that might deploy layers of consciousness might deploy complex emotions such as those named "social emotions".

4.9 Alertness, Arousal and Action Readiness

Focusing on responses and on the emotional behavior associated to emotions (such as *fear*), it can be realized a general objective of *personal protection* that integrates a set of subsequent objectives such as facing the different issues of the situation. Under a normal circumstance of fear (not caused entirely by any psychological condition), the very origin of this fear is some type of change that regards the environment.

When it occurs, the normal situation of life systems is that they perform an improved perception, in order to become more aware about the situational state of the environment. So that, artificial systems should implement the capability of to adapt themselves in this sense and the problem to face is how. Focusing on the current explanation, it will be assumed that the environment will constitute the initial cause and reason because emotion of fear emerges.

Damasio argues this displacement from the normal work as a discontinuity related to "the melodic line of background emotion" [54]. Regarding this discontinuity, what is of our interest is the viewpoint of Damasio concerning the essential processes involved. And there will be analyzed from the theoretical perspective of Damasio, associated to the study of how an artificial system might deploy the same feature.

I. Arousal Damasio defines this term under the following perspective:

> The term arousal is easier to define. It denotes the presence of signs of autonomic nervous system activation (...) which are reasonably covered by lay terms such as 'excitement'
>
> Damasio [54]

And he makes a clear distinction between this *state of arousal* from those different states of *alertness* or *readiness*. Clearly, *arousal* will be assumed as features of quite different nature that those of life systems. And each punctual design should study the requirement of this concrete type of processes within an artificial system. However, it is essential its conceptualization in order to obtain a complete portrait of artificial emotion.

II. Alertness and action readiness Damasio argues that the proper meaning of *alertness* is that situation of the agent under which it is apparently disposed to perceive and act [54]. This way, artificial systems should implement these two essential processes in order to properly behave under some emotional states.

Once a system is aware about some relevant change, it should trigger processes focused on to improve attention. But this is not the only essential capability that life systems commonly perform. The ability of survival is argued to be supported by an essential functionality of *action readiness* (see Pérez et al. [203]; Fridja et al. [79] among others). And this essential functionality is intrinsically related to the function of *alertness*.

When we refer to artificial systems, *alertness* and *action readiness* becomes the need of reconfiguring the normal work of the system to another of higher attention and preparation for action. This depends on to be aware of the current context regarding current needs of the system, deal with this context, and reconfigure the space of the problem for success.

> **Thesis Rationale ♣ 61** *Computer Action Readiness.*
> It is essentially required that system has knowledge about environment, but also perception about a situation or fact of relevance. The system must be aware of the dangers surrounding the operational space, and the situations in which they are active. On computational systems, this thesis relates action readiness with the fields of *Context Awareness* and *Systems Reconfiguration*. And it is suggested these two perspectives as the grounds to build artificial *alertness* and *action readiness*.

4.10 Discussion

This chapter shows a critical thinking about the main aspects of artificial emotion. The main objective of this chapter has been to describe the whole process of analysis made along the study of this thesis, in order to support a final model. And it has been focused on deciding what are the main ideas and theories in order to support a final design.

The ideas showed herein this chapter, involves questions about emotions that need to be asked focusing on artificial systems. There are questions that need to be reasoned from different perspectives in order to obtain acceptable responses to deal with artificial systems.

All reasonings made on the basis of this study, have been showed along the text as framed text. Since these reasonings are the logical basis behind the later model, they have been referred as *rationales* in this text.

Following chapter will describe a solution on the basis of all the rationales herein argued. All rationales will be referenced every time it would be necessary in order to support all decisions regarding each proposed approach.

Chapter 5
A Model of Computational Emotion

This thesis is concerned with the design of a reference model for emotion in artificial systems. Among the many challenges faced by this project, the comprehension of the real phenomena is one of the main prerequisites but not the only one. And in-depth analysis of existing models has been performed in order to choose those that might provide the optimal approach to each process.

Chapter 2 addresses the problems faced in choosing among the existing models and shows the perspective used in this thesis. The introduction of this topic (i.e. the conceptualization of model) was prioritized in order to bound the extent to which this work uses some terminology. Knowledge about the pure extension of solutions within the science space has been described in Chap. 3, the state of the art. However, it is in Chap. 4 that the range of processes which extend the real phenomenon is described. A study has been conducted, by moving from theory to theory, beyond the emotion science.

As a result, Chap. 4 provided those principles and foundations that will sustain the approaches that this current chapter describes. Chapter 5 thus provides a generic solution that wraps the amount of theoretical insights highlighted in Chap. 4. And this generic solution will draw a reference model available for general use, whatever the system platform might be.

Thus, the blueprint described in this chapter will not provide a concrete solution, but the generic modules of a concrete artificial system. Once specified, these modules will then conform the system capable of performing computational emotion.

5.1 Introduction

Common models of artificial emotion have stated the meaning of artificial emotion within isolated parts of the phenomena. This work aims at joining essential theories in emotion science that, once studied, have been considered as good providers of

M.G. Sánchez-Escribano, *Engineering Computational Emotion—A Reference Model for Emotion in Artificial Systems*, Cognitive Systems Monographs 33,
DOI 10.1007/978-3-319-59430-9_5

explanation for a concrete process. This chapter aims to describe a reference model through which to provide required principles and foundations for computational emotion by design. Solutions approached in this chapter are completely supported by the ideas and insights obtained in Chap. 4. A reference to each of these insights will be provided whenever it is necessary in order to clarify some conclusion.

This thesis has been developed within the framework of the *ASys Project* and with the goal of autonomous systems design. As already stated, the wide objective of this project is the improving of control strategies of systems by integrating modules of reasoning and intelligence. And those principles under which this intelligence is pursued are: (a) *Model-based cognition* that regards the exploitation of models, (b) *model isomorphism* that regards the dependence that cognitive thinking has on models, (c) *anticipatory behavior*, (d) *cognitive action generation*, (e) *model-driven perception* that understands perception as a continuous updating of integrated models, (f) *system attention*, and finally (f) the complex concepts of *system awareness* and *system consiousnes* still unsolved [229].

This chapter aims to use models in order to supply a simplified description of an emotional system to assist the performance of computational emotion, and the subsequent emergence of artificial emotion. As previously introduced in Chap. 2, the wide problem which constitutes artificial emotion requires a large number of models of quite different nature. The three major problems to model are (a) those processes that are responsible of assessing and that afterwards become the ground for experience, (b) those processes that—once obtained the ground for experience, provide the system with the observation of feeling, and (c) the self-adaptive capabilities of the artificial system to become adjusted to the new conditions that the observed situational state imposes.

Regarding the former problem, the requirement of a finite *carrier process* is conceived in order to provide a continuous emotional assessment (see Chap. 2 for reasonings regarding the finite characterization). Additionally, it seems as if the emotion based objectives, those that drive the emotional behavior, are closely related with the evolution of the emotional process. So that, it seems as if emotional drives are related to the feature of *finiteness* through the stability of the emotion-like goals.

An additional problem for discussion is the dynamic nature of the emotional phenomena, and how this dynamics can be integrated as part of the computational process. Since dynamics emerge from unstable states regarding emotional goals in life systems, they can be assumed as part of these emotional goals in artificial systems.

The issues of perception and models of experience and self are critical problems in the general challenge of artificial emotion. Since this is a long term attempt in artificial intelligence, this work assumes the requirement of relational entities in order to allow for the integration of experience in models. We hold that this relational knowledge will provide sustainments for later interpretations regarding the relation between the environment and the system. That is, regarding the effects of the environment over the system.

Tools for explanation This chapter focuses on the description of a reference model under the concluded principles expressed in Chap. 4. With the goal of a future imple-

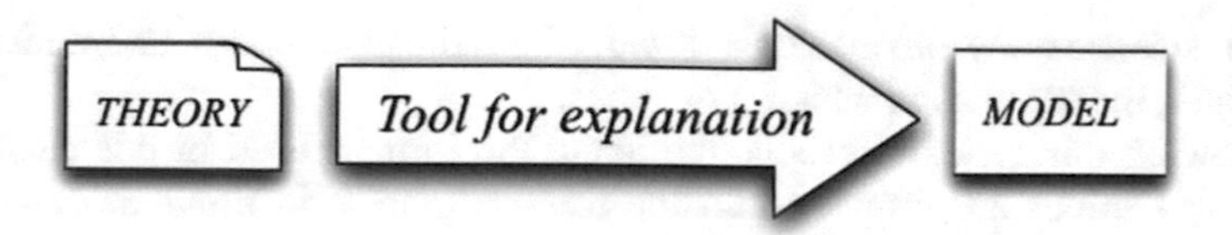

Fig. 5.1 Tools for explanation are required in order to build correspondences between theories and approaches

mentation in mind, a properly correspondence between the proposal and the theory that supports the proposed conceptual model is essential.

The final approach has been mainly performed on the basis of the entire acquired knowledge, and described within the fields of Artificial Intelligence, Model-Based Systems (MBS), and additional mathematical formalizations to provide punctual understanding in those cases that have required it. However, the main purpose of this chapter is to provide a useful description and interpretation of how computational emotion might work, and the explanation of the different foundations of the phenomenon. This model is not claimed to be a complete description of the reality about the emotion phenomena. It aims to fit a sufficient part of all the explanatory theories in the state of the art well enough, so that the approach might become useful to deploy emotion in artificial systems.

Descriptions in this chapter focus on a later subsequent implementation, so they require to be understandable and useful. To this end, different tools have been used depending on the purposes of the model which the explanation concerns (Fig. 5.1):

1. *Systems Algebra*
   The use of systems algebra under the conceptualization of Wang [267] has been aimed to provide an abstract use of the system and its elements. Essential explanations of our theory require references to specific parts inside the systems, and this tool helps us for explanatory purposes.
2. *Feedback Loops*
   This type of engines are broadly used in engineering for a large amount of adaptive-based purposes. The ones proposed by Vogel and Giese have been largely used because they are entirely defined and formalized, and this allows for a clearer explanation.
3. *Agents*
   They related terminology and some basis for illustration from the field of Artificial Intelligence have been very useful, especially considering the works of Russell and Norvig.
4. *Systems Modeling Language*
   Regarding specific explanations related to software, Systems Modeling Language (OMG SysML) version 1.2 (see SysML [254]) has been used.

Structure of the chapter Every part and structure in the proposed model has been chosen by means of inductive reasoning, on the basis of the study of the phenomena.

Emotion science splits into manageable parts the entire phenomena, and studies every case within the contexts in which it operates.

Following this general introduction about the formats used in this chapter, the following sections are grouped into the following areas in order to describe the complete problem of the emotional phenomenon.

1. *Introduction*
   This is a preliminary and explanatory section, intended to introduce the objectives and means of this chapter. This current section shows the thread of the chapter that will be followed through its presentation.
2. *Previous works*
   This section presents a summary of our previous perspectives regarding the problem this thesis addresses, in order to provide comprehension regarding how the main ideas have evolved over time.
3. *Thesis*
   This section presents a short description of the core contribution of this thesis, aiming to provide a general portrait of the approach.
4. *Emotional agent AGSys*
   This part focuses on the question of what the essentials as well as the main structure of AGSys are, in order to perform computational emotion. The conceptualization of *ESys*, questions regarding how to deal with the external environment, the abstraction of the inner environment and the interaction between these environments inside *AGSys* are the main issues addressed here.
5. *Principles of perception in AGSys*
   It is assumed that perception is intrinsically related to the ability of *AGSys* to obtain, interpret, and understand, all the information that comes from sensors. When talking about emotion, this understanding includes relevance, a concept closely connected to this work. This section provides the abstraction of the *inner-Object* as the exploitable knowledge related to an external entity, the process by which *AGSys* might approach perception, the principles of emotion perception, and the baselines for approaching the emotion representation.
6. *The inner agent ESys*
   ESys is conceived as embodied into—and living inside of—*AGSys*, and aimed to perform the complex transversal work of emotion assessment. The requirements in order to approach the appraisal work, the conceptualization made on how emotional goals might be conceived in order to provide the means for complex appraisals, and how valenced reaction might be comprehended are the main topics described in this section.
7. *Feeling in AGSys*
   The essential problem in engineering is the nature of the emotion representation, and how artificial systems might become aware of their own emotions. The problem of feeling is intrinsically joined to the problem of symbolic representation and it is described by following the perspective of *AGSys*. This problem is conceived under the construction of an *inner-Object* that grows in complexity, by means of the integration of new information related to the interaction *AGSys-object*. This

work is viewed under the operation of a recursive loop (*R-Loop*) and a model of emotion perception (*MEP*), that build an object full of patterns of data which will be finally interpreted by an emotion-based ontology. Additionally, this section objectifies our viewpoint regarding how this representation affects the autonomy in *AGSys*.

8. *Architectural Foundations*
   Once analyzed the assessment itself, the following question is how this assessment is structured within the environment of interaction *AGSys–ESys*. This reciprocal interaction and influence is an essential need for the particular purpose of emotion performance. This section analyzes some questions related to the environments of *AGSys*, the interaction and reciprocal actions between *AGSys* and *ESys*, essential questions regarding the appraisal performance and reconfiguration in *AGSys*, and an earlier idea of generalization for this approach. Additionally, other questions that regard agency and how it has been comprehended in this work have been analyzed.

5.2 Previous Works

Previous works that we developed related to this thesis (see Sánchez-Escribano and Sanz [227]), approached the problem with an organization in terms of two subsystems that concurrently managed the external domain, and the created internal domain (see Fig. 5.2). From the same perspective adopted in this thesis, the interaction of the longitudinal system with the external domain, and the interaction of the transversal system with the inner one were conceived.

The main objective of the transversal system aimed to apply feedback from the inner domain of the longitudinal system over the system itself. This way, it might be able to perform suitable decision-making concerning the value-based meaning of the inner-domain state.

The longitudinal system was conceived as a collection of distributed system-of-systems (SoSs) working federated to cope with general system objectives. The system as a whole should fulfill the extrinsic goals, while maintaining the success of the intrinsic goals. The extrinsic and intrinsic goals were conceived as the targets related to the system mission and system wellness respectively.

The transversal system was devised as an orthogonal system which monitors the inner domain of the longitudinal system. This inner domain is assumed to be full of changes occurring inside the longitudinal system. So that, it is a provider of an environment full of data aimed to obtain exploitable models for later actions.

Later in the work of this thesis, new requirements related to models, integration of theories, actions and generalizations emerged. So that, earlier conceptualizations have evolved in order to provide the required artifacts for our approach.

This way, the longitudinal system has become a rational agent *AGSys* that works within the framework of a system—whatever its final form. And the work of the transversal system has been conceptualized under the operation of *ESys*. The inner

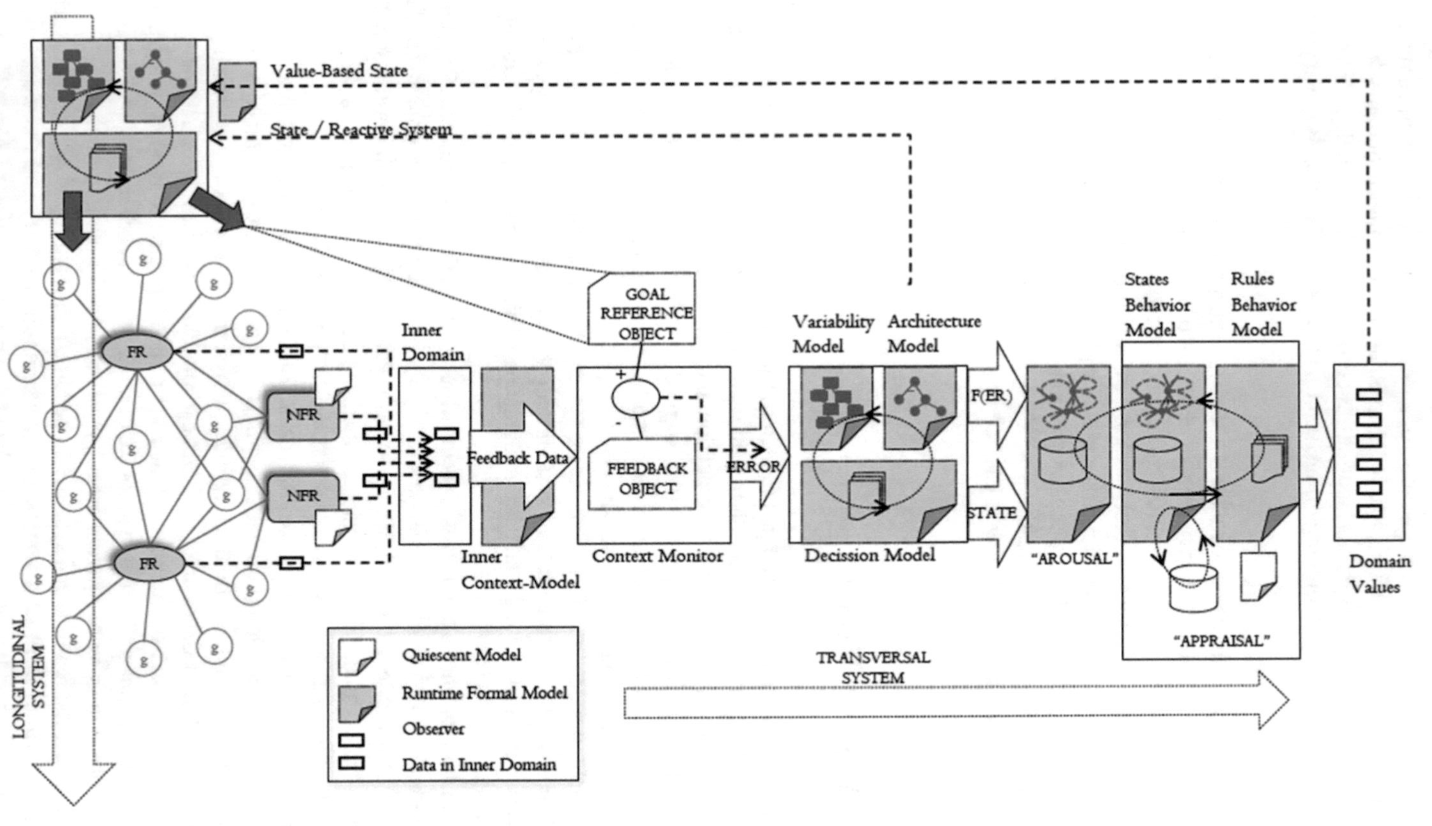

Fig. 5.2 General view of a previous conceptualization regarding the ideas of this thesis

goals are defined as a structure of an agent-based model, and monitored under the operation of reflex-agents. And the external goals are comprehended from the perspective of common targets related to the system's mission, without the requirement of being specifically denoted for the purposes of this thesis.

5.3 Thesis

A general overview of this thesis is illustrated in Fig. 5.3. It describes a reference model to feedback systems with valuable meaning, allowing for reasoning with regard to (a) the relationship between the environment and the relevance of the effects on the system, and (b) dynamic evaluations concerning the inner situational state of the system (as a result of these effects). It consists of a multi-purpose architecture that implements two broad modules in order to operate: (a) the range of processes related to the environment affectation, and (b) the range or processes related to the emotion perception-like and the higher levels of reasoning.

The problem has been interpreted and is described on the basis of *AGSys*, an agent assumed to have the minimum rationality to provide the capability to perform emotional assessment. *AGSys* is a conceptualization of a *Model-based Cognitive agent* that embodies an inner agent *ESys*, the responsible of performing the emotional operation inside *ASys*. Throughout this solution, the environment and the effects that might have an influence on the system are described as different problems.

While *AGSys* operates as a common system within the external environment, *ESys* is designed to operate within an *inner environment* built on the basis of those relevances that might occur inside *AGSys*. This allows for a high-quality separated reasoning concerning mission goals defined in *AGSys*, and emotional goals defined

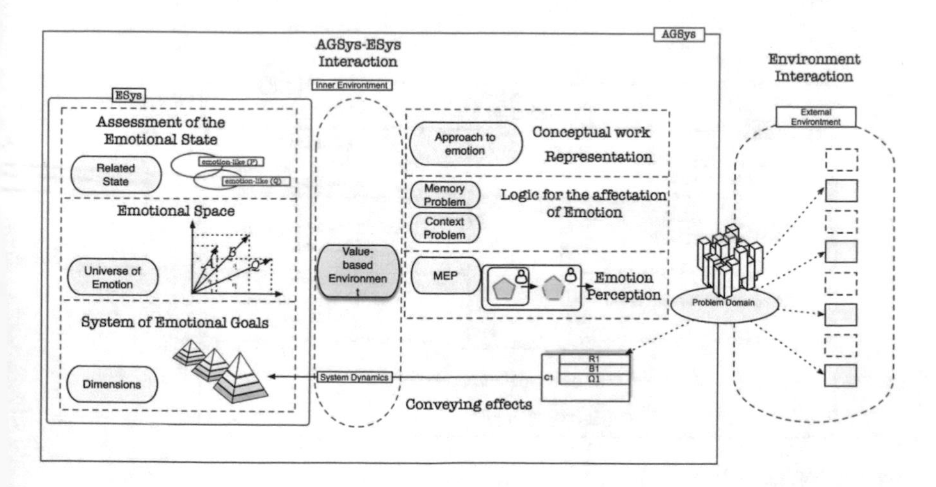

Fig. 5.3 Thesis

in *ESys*. This way, a possible path for high-level reasoning under the influence of goals congruence is provided.

The high-level reasoning model uses knowledge about emotional goals stability, hence allowing for new directions in which mission goals might be assessed according to the situational state of this stability. This high-level reasoning is grounded by the work of *MEP*, a model of emotion perception that is conceived as an analogy of a well known theory in emotion science. The work of this model is described under the operation of a recursive-like process denoted *R-Loop*, together with a system of emotional goals that are assumed to be individual agents. This way, *AGSys* integrates knowledge about the relation between a perceived object, and the effect which this perception induces on the situational state of the emotional goals. This knowledge enables a high-order system of information that provides the sustain for a high-level reasoning. The extent to which this reasoning might be approached is just delineated and assumed as future work.

Principles of design The problem of emotion is conceived from the study of several broad problems that require specific analysis and design: (1) The dynamic of errors related to the emotional goals, (2) the process of emotional assessment, (3) the emotion perception, (4) the approach to emotion representation, and (4) essential foundations of design, in order to enable subsequent requests for functional reconfiguration. So that, this thesis studies the problem focusing on the following questions, as illustrated in Fig. 5.4:

1. Model of Computational Emotion in ESys
   (a) The design of the emotional goals in such a way that allows to analyze the complex behavior in their interaction. Emotion is caused by this interaction rather than by specific errors of accomplishment.
   (b) The integration of dynamics related to the accomplishment of emotional goals.

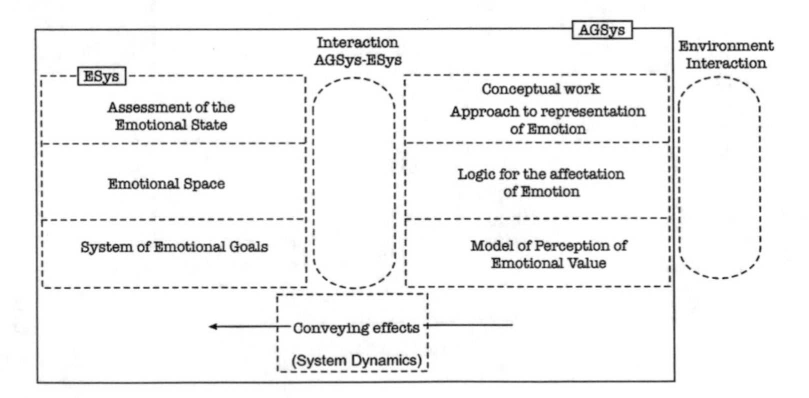

Fig. 5.4 Thesis

(c) The design of an emotion appraisal space, within which the situational state of emotional goals can be analyzed.
(d) The design of a methodology aimed to obtain the final assessment, on the basis of the situational state of the emotion appraisal space.

2. Operational Model for Interaction ESys–AGSys

(a) The design of an abstract inner environment as the framework where the accomplishment of emotional goals is monitored.
(b) The design of an interaction between ESys and AGSys, aimed to enable functional reconfiguration in AGSys and emotional modulation in ESys.

3. Model of Emotion in AGSys

(a) The problem of how the emotional assessment is integrated as part of an exploitable model.
(b) The design of a logic aimed to allow for context reconfiguration, memory recalls, cognition and behavior.
(c) The question of how a representation of the final emotion, in order to allow the system for the exploitation of this knowledge, can be approached.

5.4 Emotional Agent AGSys

An emotional agent is conceived as an agent able to deploy artificial emotion, and it results in the conception of an agent able to perform computational emotion. Now, the question is what the essentials and the main structure of AGSys are, in order to perform computational emotion.

As explained in Chap. 3, *AGSys (Agent Generic System)* is conceived as a generic agent which might deploy artificial emotion. Following the argument of Russell and Norvig, the body architecture of this agent is considered as either a physical or computer platform. The only requirement of this generic agent is an agent program that allows to perform some sort of rationality under the conception of Russell and Norvig.[1]

The idea that regards how much the arrangement of *AGSys* will depend on the *agent program* will be accepted. The model-based approach to cognition in *ASys Project* (together with the accepted perspective of *cognitive approach of emotion*), results in the conceptualization of *AGSys* as a *cognitive agent* which implements a *cognitive architecture*.

As previously justified in Chap. 2, software architectures provide larger capabilities of abstraction. Consequently the *agent program* will be assumed as supported by a cognitive architecture, which herein will be assumed generically as a software architecture which integrates cognitive-like abilities:

[1]It is able to (a) performance measurements, (b) use prior knowledge, (c) behave based on knowledge, and (d) store and use the sequence of perceptions till the current time.

$$AGSys = (\text{Body Architecture})_{(AGSys)} + (\text{Software Architecture})_{(AGSys)} \quad (5.1)$$

This dissertation aims for objectives of generality, and it imposes a requirement of (Body Architecture) independence by design. Nevertheless, it will be assumed that (a) (Software Architecture) and (b) (Body Architecture) will follow such descriptions and forms as to perform with success regarding the requirements of environmental perception (i.e. sensors) and action (i.e. actuators). This results in the essential requirement of connectivity between the solution approached in this dissertation, and the final system where it might be hosted. And this brings about the future need of modeling the ad-hoc bridge between the final platform and the solution proposed in this thesis.

Regarding *AGSys*, it still does not have a concrete description in order to provide a conceptual basis for explanatory purposes. However, since *AGSys* is argued to be a cognitive agent focused on principles of high order rationale, it will be assumed to be a complex abstract system. And the basis for both conceptualizations -*AGSys as a complex system* and *AGSys as an abstract system*- will be provided.

I. AGSys as a complex system Complex systems are large systems that involve substantial amounts of components, or systems which implement different levels of abstraction and reasoning, complex behaviors, complex restrictions, etc. From the perspective of Magee and de Weck, complex systems are distributed *systems of systems (SoS)* composed by interrelated parts that perform a federated work [157].

Since an essential foundation in *cognitive systems* is the interaction with the environment in order to exploit models [229], *AGSys* can be devised as an *open system* under the conception advanced by Von Bertalanffy et al. [265]. That is, a system that maintains itself in exchange of materials with the environment, and is continuously building up and breaking down its components [265].

This can be also assumed to be an essential feature regarding emotional modeling. Since the real phenomena of emotion is framed into life systems, it is essential to take into account the viewpoint of Von Bertalanffy et al. that assumes "the characteristic state of the living organism is that of an open system" (since emotion seems to be justified under the circumstances of an open system) (Fig. 5.5).

II. AGSys as an abstract system

> Systems are the most complicated entities and phenomena in the physical, information, and social worlds across all science and engineering disciplines. Systems are needed because the physical and/or cognitive power of an individual component or person is not enough to carry out a work or solving a problem
>
> Wang [267]

Wang argues that the generic theory of systems tends to trait everything as a system, understanding complexity under the idea of systems that contain subsystems and that pertain to other [larger] systems [267]. Regarding the objectives of this thesis, the essential idea is that systems, subsystems or any type of component within the system have continuous interaction.

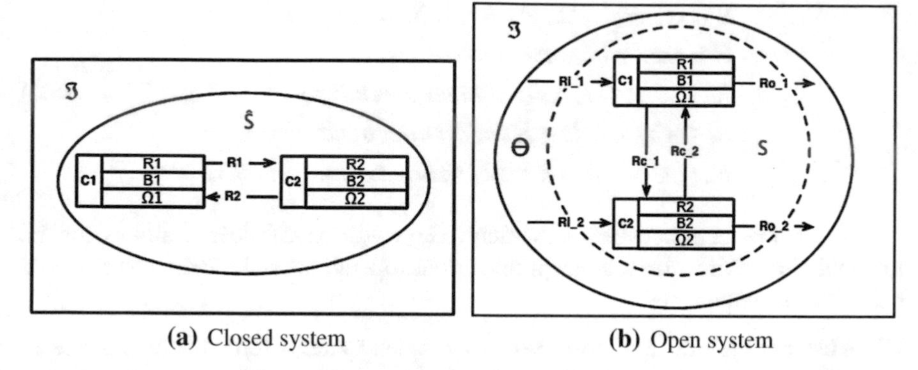

(a) Closed system **(b)** Open system

Fig. 5.5 The abstract model of an open and a closed system from the point of view of Wang (adaptation from Wang [267])

Thus, whatever the general idea of a system, it intrinsically involves the concept of interaction between *system components*. The problem that regards this dissertation, is how to describe this concept of *interaction* under a successful formalization that can be properly used.

Wang shows in his work "On Abstract Systems and System Algebra" a formal way to rigorously trait a system as a mathematical entity. And he introduces the concept of "abstract system" as well as those mathematical models that constitute it. He assumes system algebra as the foundation for this conceptualization, arguing that it provides a formal treatment of abstract systems and their properties, and—by paragraphing the words of the author—allows for "manipulating system operations and their composition rules".

> *Definition 1.* An abstract system is a collection of phenomena in the physical, information, and social worlds coherent and interactive entities that has stable functions across all science and engineering disciplines. Systems have a clear boundary with the external environment.
>
> *Definition 2.* A system algebra is a new abstract mathematical structure that provides an algebraic treatment of abstract systems as well as their relations and operational rules for forming complex systems
>
> Wang [267]

Wang defines a closed system as a 4-tuple of components (C), relationships (R), behaviors or functions (B) and constraints (Ω):

$$\hat{S} = (C, R, B, \Omega) \tag{5.2}$$

An open system is defined as a 7-tuple that integrates the environment (Θ) and the relations with the environment (i.e. inputs and outputs), together with the previous four components:

$$
\begin{aligned}
&S = (C, R, B, \Omega, \Theta) = (C, R_c, R_i, R_o, B, \Omega, \Theta)\\
&\Theta \textit{ is the environment of the system } S\\
&R_c \subseteq C \times C \textit{ is a set of internal relations}\\
&R_i \subseteq C_\Theta \times C \textit{ is a set of external input relations}\\
&R_o \subseteq C \times C_\Theta \textit{ is a set of external output relations}
\end{aligned}
\tag{5.3}
$$

So that, this perspective from which *AGSys* will be considered allows for the manipulation of properties, conceptualizations, operations and relationships in, and between its components.

III. Abstract systems and complexity Complex systems force the emergence of new effects with no direct causality related to any defined component. The interaction between the complex system and its environment produces the emergence of additional effects inside the system that constitute a complex set of relevant new circumstances that are also influencing the system.

The work of Wang allows for mathematical models of abstract systems and formalization of the relations between these abstract systems but this is not the only advantage. It also allows for the manipulation of these systems and relations by means of an algebraic system (i.e. the system algebra).

This conceptualization offers a relevant perspective for this dissertation, since it provides a means to explain how to capture essential relationships within a complex system.

5.4.1 Conceptualization of ESys

One essential argument in this work is the assumption of emotion as a transversal functionality of life systems (see Sanz et al. [231]), and this assumption directly affects the *Software Architecture* of *AGSys*. This *Software Architecture* will be considered as the support for two type of works: (a) the one related to the rationale of *AGSys* (under the normal conception of rationale in artificial intelligence), and (b) a transversal work in order to attend the transversal functionality of emotion (Fig. 5.6).

As introduced in Chap. 2, this idea is conceived by decoupling the external and the inner environments of *AGSys*. The conceptualization of transversality is also influenced by the idea that emotions work inside systems—in order to maintain a systemic equilibrium concerning system wellness. Somehow, this idea derives toward the subsequent suggestion of an inner environment where an *agent-like emotion* lives.

Thus, a new agent named *ESys (Emotional System)* embodied into and living within the inner environment of *AGSys* will be conceptualized. This way, the *utility* (conceived as an attribute of this new agent), the *performance measurement*, and the *utility function* of *ESys*, will be related to sequences of those inner situational states of *AGSys* rather than to those sequences related with its external environment.

Finally and for the sake of preventing ambiguities, this work will refer to the complete phenomena realized in life systems (i.e. in the sense of 'a system that

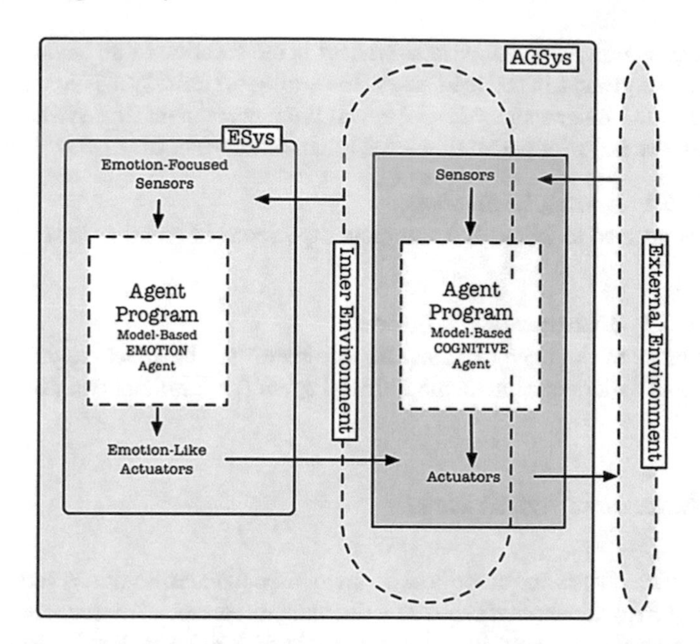

Fig. 5.6 Essentials and main structures of the emotional AGSys according to the approach proposed in this work, in order to perform computational emotion

deploys artificial emotion'). That is, *artificial emotion*. That is, *artificial emotion* will be differentiated from *computational emotion* in the sense that the former will refer to the complete phenomena that an artificial system might deploy, and the latter will just refer to the computational work that will sustain the former (see Fig. 5.7).

Under the conceptualization of *AGSys* as a cognitive agent under the form described in Eq. 5.1, and the conceptualization of a new agent *ESys (Emotional System)* embodied into and living within the inner environment of *AGSys*, an interaction between *AGSys* and *ESys* is assumed in order to build the *inner environment*.

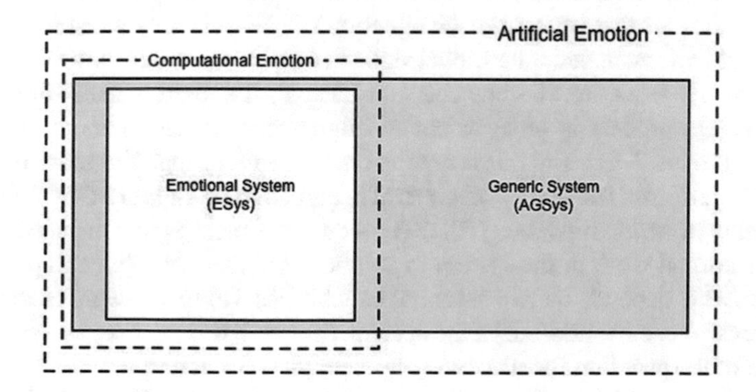

Fig. 5.7 Conceptualization of Artificial Emotion and Computational Emotion (see rationale ♣ 7)

The justification for an *inner environment* is the creation of an environment that will serve as a ground in order to assess the wellness of *AGSys*. This wellness will be related to the inner states of *AGSys*, and the environment of *ESys* has to be the one that allows *ESys* for perceiving what might be relevant for *AGSys*.

Definition 5.1 ♣ *Artificial Emotion*
General name used to denote the complete experience of emotion in some artificial agent.

Definition 5.2 ♣ *Computational Emotion*
Computational work aimed to obtain the complete set of artifacts required to build the emotional experience into some artificial agent (i.e. artificial emotion).

5.4.2 External Environment

Focusing on responses and emotional behavior associated to emotions such as *fear*, a general objective of *personal protection* that integrates a set of subsequent objectives such as facing the different issues of the situation can be realized. Under normal circumstances of fear not caused entirely by a special psychological condition, the very origin of fear is some type of change regarding the environment.

When it occurs, the normal situation of life systems is that they perform an improved perception, in order to become more aware about the situational state of the environment. So that, artificial systems should implement the capability of adapting themselves in this sense. The problem now is the relationship between *AGSys* and the environment within which it performs its mission. So that, in order to analyze this issue, there is no essential requirement to analyze complex processes of emotion emergence regarding fear. For now, we will assume that the environment will constitute the initial cause and reason why emotion emerges.

The environment of any biological system is extremely complex to be dealt with as a complete problem. Life systems are not aware of the complete environment, but only about the part that concerns each situational state. It is only when it is required, that systems increment their attentional skills in order to complete partial knowledge.

Once *AGSys* is aware of some relevant change, it should trigger some type of *attentional-like* process in order to obtain improved information about this change (see thesis proposal 61). But this is not the only essential capability that life systems commonly perform. The ability of survival is argued to be supported by an essential functionality of *action readiness* [79, 203]. And this translates into the need to reconfigure the normal work of the system to another of higher alertness and preparation for action. This depends on two essential capabilities: (a) to be aware of the current context regarding current needs of the system, deal with this context, and reconfigure the space of the problem for success, (b) preparation for action.

The emotion of fear will be used in order to illustrate the general approach defended by this thesis. First of all, the conceptualization of *fear* is required from the perspective of computer science, in the form of an *event*. That is, without allusion to any process that might define the emotion. The only requirement is an experience that something of importance is happening.

From the perspective of a life system, the concept of *fear* exists through its life, and it acquires different forms and values depending on the situational state of the environment. The analogous problem will be illustrated from the perspective of computer science (since this is the science that deals with the program of the agent). Now *fear* is an abstract concept, and this concept maintains a continuity through the life cycle of the system and passes through multiple forms. In the field of computer science, the conceptualization associated closely with this abstraction is the abstract idea of *entity* (see ENTITIES by Evans [74]).

During a normal mission, *AGSys* works within a *sub-domain* of its domain. That is, *AGSys* works within a delimited part of its entire problem space: the one that refers to its mission. Computationally speaking, *AGSys* implements a *domain model* as a software artifact that defines the solution space. During a normal mission, *AGSys* is working within a *bounded context*, the software artifact that derives from the *domain model* and that delimits its applicability. This way, the requirement to change this *bounded context* when *fear* appears can be assumed. Assuming that *AGSys domain model* integrates a *bounded context for fear*, reconfiguring the previous *bounded context* to this for fear attention is required (see Fig. 5.8).

Generally speaking, *context* refers to any information that can be used to characterize the situation of any entity (see Dey et al. [62]). According to the perspective

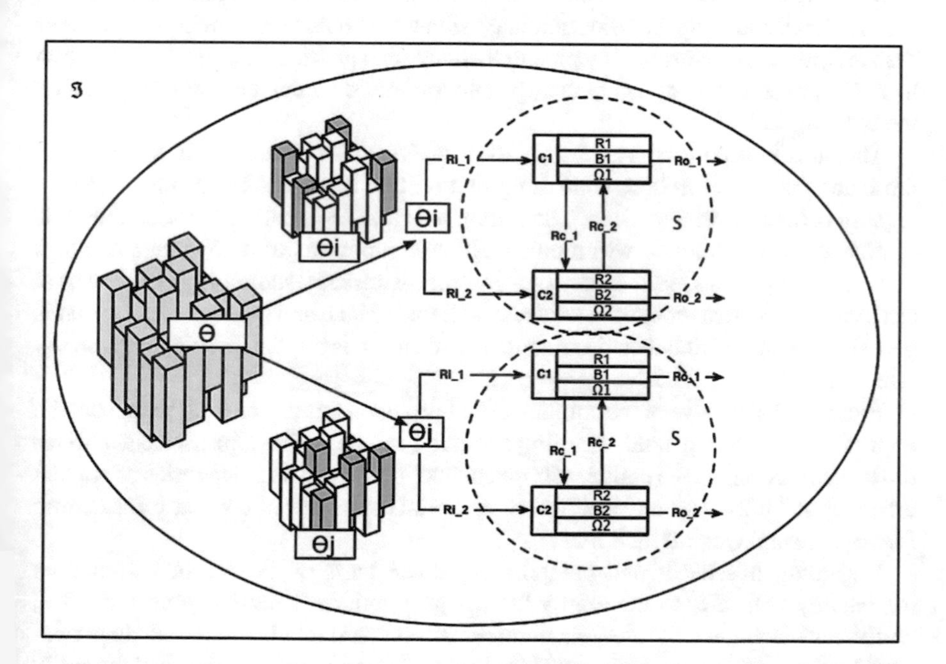

Fig. 5.8 Conceptualization of environment reconfiguration: Θ_i represents the normal work of *AGSys*, and Θ_j the one related to an environment that conveys the situational state of fear to *AGSys*

of Dey et al., the environment has three main parts: computing environment, user environment and physical environment. Regarding *AGSys*, the physical environment will depend on the ad-hoc platform that it might implement. The user environment represents the interaction with external agents, from which the event that might cause the emotion would come. And the computing environment will be the one that refers to the reconfiguration for *action readiness* and *attentional-like* abilities.

As introduced in the thesis proposal 61, the challenge regards the field of *context-awareness*. That is, how to change and further improve the quality of *AGSys'* behavior by adapting its operation to the situational state of the environment. The problem is twofold: (a) *AGSys* has to know the change of the environment, and (b) *AGSys* has to reconfigure itself in order to attend to that situation.

A clarification regarding the meaning of 'to attend that situation' is herein required. An emotional agent triggers self defensive behaviors even when it does not completely recognize a situational state of the environment. This is the real ability that improves the protection capability of emotional life systems. Analogously, it will not be necessary for *AGSys* to have a complete knowledge about a situation to attend to it with objectives of self protection. This provides improved system behavior for any sort of technical system, as argued in previous works (see Sánchez-Escribano and Sanz [227]).

The main goal is to understand and manage a situation, even when this situation is unknown. The emotional capability allows for understanding not the situation (what would require the complete knowledge of that situation), but the meaning of that situation, faster, in order to behave safely. The proposal that this thesis makes is to focus on understanding the "emotion like" situational state of the environment rather than the environment itself. Any punctual change can provide emotional information to *AGSys*, even if this punctual change cannot provide complete knowledge about the environment.

This idea is supported by the operation of *ESys*. This inner agent will monitor the inner changes of *AGSys*, something that will be explained by means of system algebra in the following section. The rationale is that *ESys* will monitor on the basis of emotional like goals. It will monitor either components of *AGSys*, relationships between components, influences, etc. instead of directly monitoring the external environment. So that, context-awareness will understand and manage context that is perceived in an emotional environment, and regard it as implicit inputs to efficiently affect the behavior of *AGSys*.

From the perspective of the computational problem, this can be dealt with from the approach that Dey et al. hold regarding context-awareness. This approach categorizes the context "based on the entities whose context is assessed and categories of context information" [62]. And introduces four essential categories of context information: *Identity*, *location*, *status*, and *time*.

Regarding this thesis and paraphrasing some parts of the author's discourse: (a) *identity* will refer to the ability "to assign a unique identifier to an [emotion] entity", (b) *location* will refer to any information that might be used to deduce relations between entities, (c) *status* will identify the "intrinsic characteristics of the entity that can be sensed" (a complex work that will be made by means of the structure of

AGSys and *ESys*), and (d) *time* as an essential feature to characterize an emotional situation.

This objective of emotion-based context modeling requires additional developments such as domain languages that allow for reconfiguration, abstraction and platform independence. This issue will be addressed later as part of this research, but a general outline of this proposal on the basis of the work made by Hoyos et al. [111] can be described here.

This work has been developed within the framework of context-awareness and the goal of allowing systems to react based on their environments. Within this framework, the problem of the context management is solved by means of a textual Domain-Specific Language (DSL), especially adapted for modeling context information. On the basis of this work by Hoyos et al., it is possible to describe how, by connecting each emotion to its relevant context, the reconfiguration under an emotional behavior can be viable for the system.

From the analysis in Chap. 4, computational emotion can be argued to be built on the ground of three essential artifacts: appraisal dimensions, concerns and goals. By linking these three artifacts to an emotion and its context, a scheme such as this can be obtained, as showed in Fig. 5.9. This is not a final proposal for *AGSys* context model, but just an illustration of how this final model might be in order to obtain a reconfigurable context for *action readiness*.

5.4.3 Inner Environment

Having *AGSys* as a generic cognitive agent in the form described in (5.1), and the conceptualization of a new agent *ESys* embodied into—and living within—the inner environment of *AGSys*, some type of interaction between *AGSys* and *ESys* is assumed. And this interaction will support the building of the inner environment.

The justification for an inner environment is the creation of a sphere of activity that will become the environment within which *ESys* operates. Emotion seems to perform autonomously concerning some essential relevances in life systems, causing a broadcast of dynamics and cues that allow them for readiness and rationale. Life systems are brain based systems, far from the operation of artificial systems. So, the replication of the original structures that perform in nature is not required but a model that allows for analogous consequences is needed.

An inner agent *ESys* has been conceived, whose rationale is driven to the accomplishment of these essential relevances that emotions regards. These relevances can be understood as transversal goals for *AGSys* in order to maintain its systemic equilibrium for wellness. As this systemic equilibrium strongly depends on the inner dynamics and components interactions within *AGSys*, the better solution is to convert these interactions and dynamics into the environment of *ESys*.

This new environment will be named *inner environment*, and has to be created on the basis of these interactions. The final purpose is the building of models that provide causal connections between the real environment of *AGSys* (i.e. the external

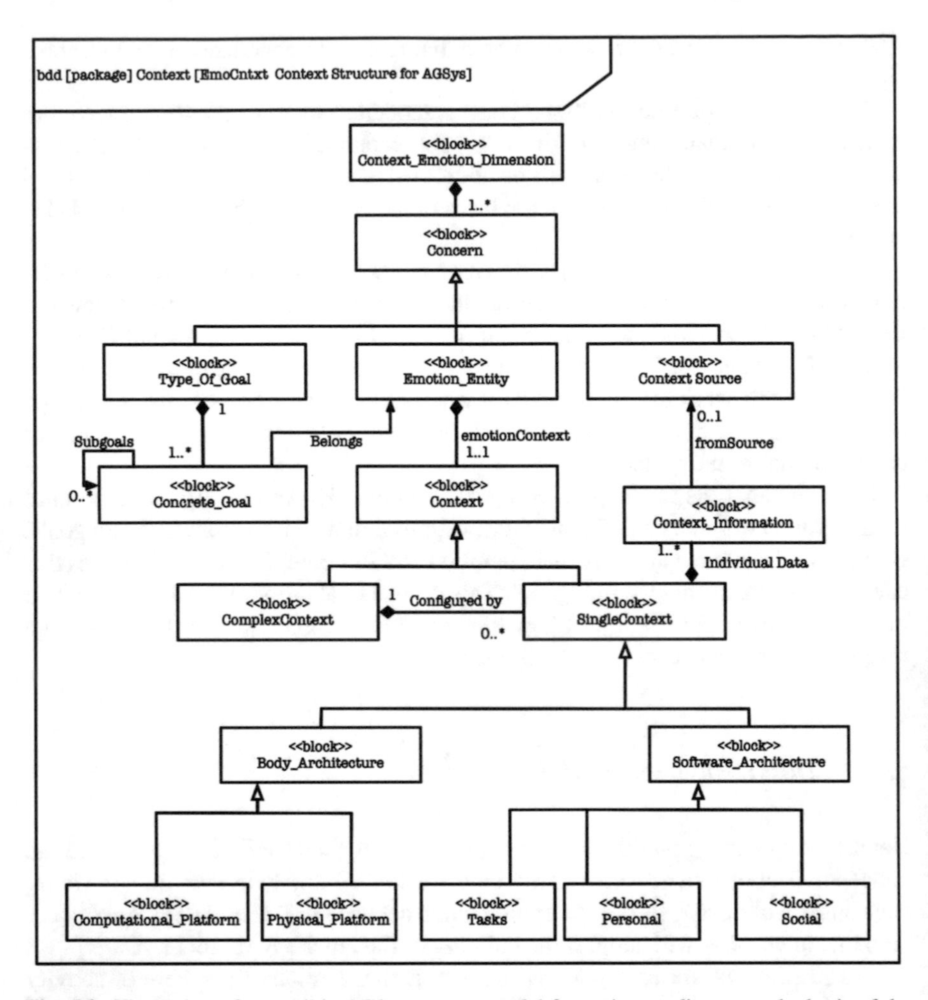

Fig. 5.9 Illustration of a possible AGSys context model for action readiness, on the basis of the MLContext's abstract syntax metamodel of Hoyos et al. [111]

environment), and the effects that this external environment has concerning its relevances for wellness. That is, a way to manage the effects of the external environment concerning the systemic equilibrium of *AGSys*.

I. Formalization of AGSys By assuming *AGSys* to be a *cognitive agent*, it can be argued that *AGSys* is a complex system with interactions with the environment by means of which exploit models. And it enables us to characterize *AGSys* as an open system according to the idea of Wang.

Therefore, the mathematical formalization of our generic agent *AGSys* can be as follows:

$$AGSys = (C, R, B, \Omega, \Theta) = (C, R_c, R_i, R_o, B, \Omega, \Theta) \tag{5.4}$$

Since this formalization allows for generalization, there is no required allusion herein about whether components (C), relations (R), behaviors (B), constraints (Ω) and environment (Θ) are related to *Body Architecture* or *Software Architecture* of *AGSys*.

II. Formalization of the inner environment The work of Wang provides a set of algebraic operations on systems that are of relevance to our explanation: conjunction, disjunction, difference, composition, and decomposition are defined. And it also provides means to formalize system properties such as coordination, synchronization, complexity, hierarchical and recursive architectures.

Emotion establishes a set of relevant dimensions under which assessments are made [71]. With the conceptualization made in this thesis, the relevance of the external environment is captured by means of a causal relationship that measures the sensitivity of the system to 'something'. This explanatory artifact of 'something' is conceived as the expected value of the monitored sensitivity under the assessment that an emotional dimension imposes. It is thus necessary to measure the sensitivity (i.e. relevant quantities within the system) determined by the relevance that imposes an emotional dimension (see proposals 41 and 42 in order to cite evidences in support of this idea).

The inner environment is thus conceived as a software-based description, in order to obtain a virtual environment inside *AGSys* that will constitute the external environment of *ESys*.

Consequently, the inner environment should be provided with artifacts able to make the following work: to measure either components, relationships, behaviors or so, with relevance for the system. Relevance will be defined under the requirements of each emotional dimension.

The artifacts that will provide measurements will be software agents that will be named Error-Reflex agents. Error-Reflex agents will monitore the inner environment of AGSys and will submit the result to ESys under the form of an error (Fig. 5.10).

These artifacts are conceived as simple reflex agents under the conception described in Chap. 2, that is, agents that act regarding the current percept with no need for the percept history. The reason for choosing agents is the reasonable argument of flexibility in design. Agents are artifacts whose programs implement the agent function, and allow for solutions from lower to higher degrees of complexity depending on ad-hoc requirements.

Definition 5.3 ♣ *Error-Reflex agent*
Software agent aimed to monitor the inner state of *AGSys* concerning the requirements imposed by some emotional dimension.

These agents are not meant to directly measure errors regarding some component, but to implement functions that allow for monitoring properties among them according to the concept of Wang (such as coordination, synchronization, complexity, etc.). It is necessary to highlighten that in case a simple error measurement might be required, this solution admits that particular implementation.

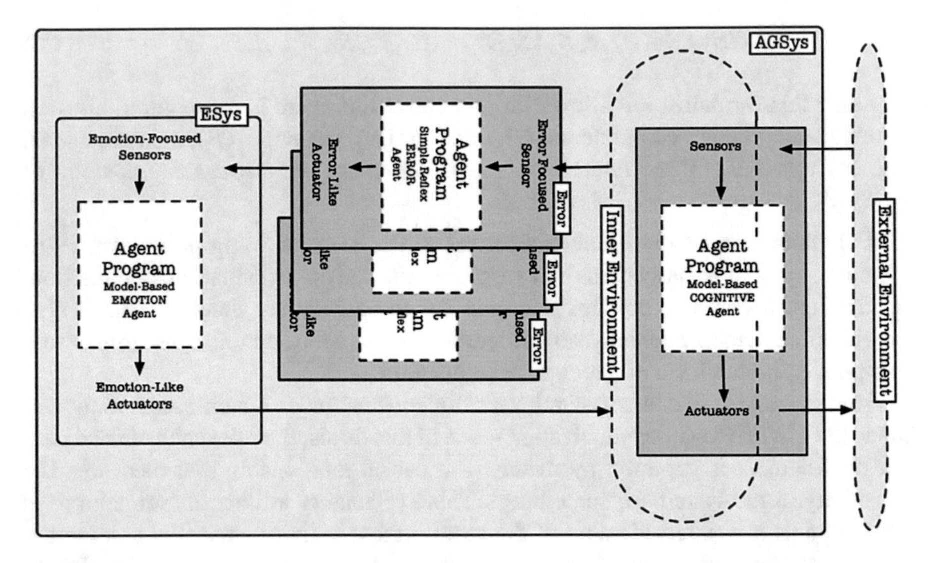

Fig. 5.10 Structure of the emotional AGSys that shows the integration of the Error-Reflex agents

▼

```
function ERROR–REFLEX AGENT (measurement) returns an error
persistent: condition–error rules
state=I NTERPRET–INPUT (measurement)
rule=R ULE-M ATCH (state, rules)
state=rule.ERROR
return error
```

▲

Fig. 5.11 A generic program for an Error-Reflex-Agent (adaptation made on the reflex agent program by Russell and Norvig [223])

Figure 5.11 shows a generic agent program in some punctual form. From this perspective, the conceptualization of inner environment can be defined in the following form:

Definition 5.4 ♣ *Inner Environment*
Software-based description in order to obtain an environment within which *ESys* may accomplish its missions, related to emotional requirements. Consequently, it will be constituted by those values that *Error-Reflex agents* provide.

III. The appraised subsystem It will denote *appraised subsystem* those parts whose interactions, states, etc. are susceptible of being analyzed under some emotional

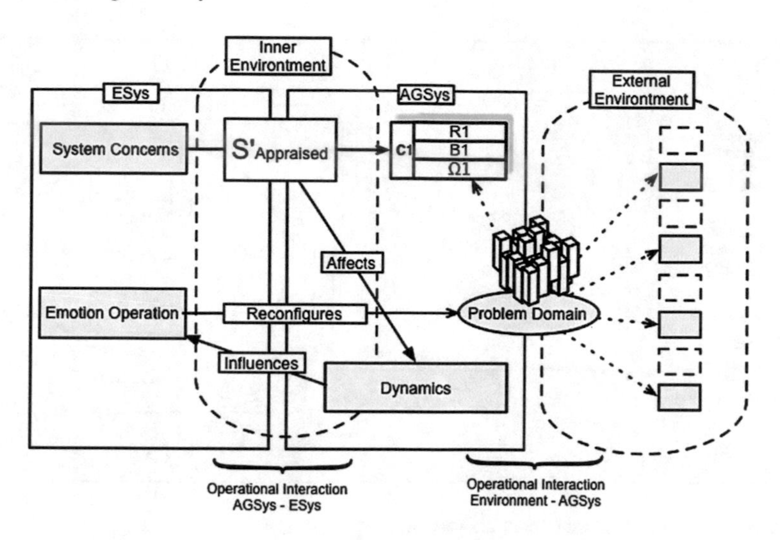

Fig. 5.12 Relative location and interactions of the Appraised Subsystem

requirement, and it is formally defined in (5.5). That is, the *appraised subsystem* is assumed to be any target area inside *AGSys* that is susceptible of being analyzed based on cross requirements for wellness (Fig. 5.12).

Definition 5.5 ♣ *Appraised Subsystem*
It is the system formed by a set of independent components of *AGSys* whose interaction might become critical under the requirements of some emotional dimension. And this appraised subsystem can be defined under the following formalization:

$$S'_{appraised}(C', R'_c, R'_i, R'_o, B', \Omega', \Theta') \subseteq S_{AGSys}(C, R_c, R_i, R_o, B, \Omega, \Theta) \quad (5.5)$$

$$S'_{appraised} = \bigsqcup_{i=1...N} S_i(C_i, R_{ci}, R_{ii}, R_{oi}, B_i, \Omega_i, \Theta_i) \quad (5.6)$$

$$\text{Where } (\bigsqcup_{i=1...N}) \text{ is related to some relevance} \quad (5.7)$$

A complete description of the inner environment conceptualization is shown in Fig. 5.13. Fundamental properties of systems such as those described by systems algebra [267] (i.e. coordination, synchronization, complexity, hierarchical and recursive architectures), are confirming grounds for emerging events at runtime, and most of the times they are not considered during the design phase of the system. And the building of resilient systems that can address their missions in a dependable way is a major challenge. Under the 4th principle of systems science, there is no system that might be efficient 100% and the resilience will not ever be assured (see Theorem 7 in Wang [267] for this principle of systems science).

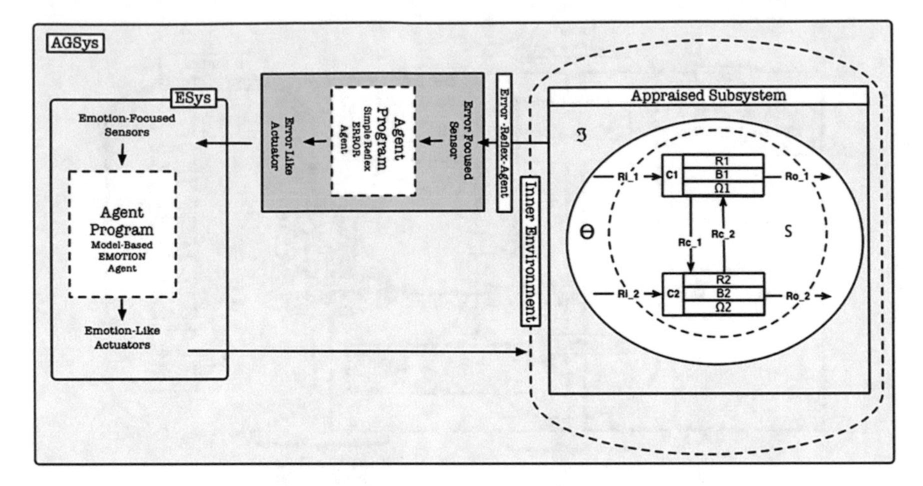

Fig. 5.13 Inner environment conceptualization

5.4.4 Ongoing Interaction Between Environments

Major relevance of an emotional transversal work is the integration of high order objectives that influence the normal work of the system. Emotion phenomena seem to be related to a complex system of concerns, which influences the work in a transversal way related to the normal operation of life systems. Whatever the mission of life systems, this mission is always influenced by this transversal work of the emotion phenomena.

On the other hand, emotions fulfill adaptive, social and motivational functions providing the most adequate behavior concerning each punctual situation. From the viewpoint of an artificial system, this is a problem of runtime environment reconfiguration regardless, by now, the urgency that this reconfiguration might require (i.e. action readiness during an emotion of fear requires faster processes than a situation related to motivation). This issues need to be attended to from different points of view.

We have described how each punctual problem might mark out the boundary (or limits) within the solution space of an artificial system. The problem is the interaction between these two environments in order to fulfill the concrete emotional process of *action readiness*.

Clearly, what is required is an emotional theory that is firm in reliability, according to which the artificial phenomena will be explained. This theory will ground those principles under which emotional concerns and environment relevances will be defined within the artificial system.

Considering the influence of the emotional concerns defined in *ESys*, the scope of subsystems that should be appraised in *AGSys* is stated. The external environment and *AGSys* are under continuous interaction that might cause effects on *AGSys* individual components.

In a general explanation: If some relevance is realized, the associated dynamics convey cues in order to trigger the reconfiguration of the problem in *AGSys* (see Fig. 5.12). These dynamics are related to the work of *ESys* concerning the emotional goals, and it will be detailed in the following sections.

5.5 Principles of Perception in AGSys

Thesis 1 ♣ Emotion vs. Perception
Emotion is not a question of *'knowing value'* but a question of *'perceiving value'*. This thesis assumes the ability to obtain value as an ability closely related to the *perception* of this value.

The arguments that support what this section assumes (regarding a model of perceptual modeling) come from the discussion about the problem made in Sect. 4.5 of Chap. 4. This section is focused on describing a possible model of perception. Focusing on emotion perception, one related essential idea is feeling. It is assumed to be an intermediate stage between emotion and the experience of emotion [54]. So that, this thesis holds that a model of perception is required (in order to build the artifact that will integrate the essential chunks of information that will allow for feeling).

It is assumed that perception is intrinsically related to the ability of *AGSys* to regard, interpret, and understand all the information that comes from sensors. When emotion is referred to, this understanding regards relevance, a concept closely connected to situations of wellness consideration. Sometimes this relevance is referred to as *value*, or the degree of importance that something has for the system.

Whatever the perspective from which relevance can be conceived, what is clear is the requirement of providing our artificial system *AGSys* with the necessary skills to perceive this relevance (see rationale ♣ 28). Perception is argued to be the responsible of measuring relationships regarding the emotional references that concern *AGSys*.

5.5.1 The Inner-Object

Since *AGSys* is conceived as a knowledge-based system capable of deploying sensory, perceptual and conceptual processes, we can argue that our system can represent *abstract objects* from the perspective of Gruber [98].

The concept of abstraction results thus in the capability of *AGSys* for building a representation of an *object*, understanding *object* in the sense of Damasio (that is, in a "broad and abstract sense—a person, a place, and a tool are objects, but so are specific pain or an emotion" [54]). That is, anything that comes from the real environment, and that is re-built by the system under a required abstraction, and in order to understand it.

Definition 5.6 ♣ *inner-Object*
Computational artifact that we situate inside of the system, which represents those descriptive aspects of the external objects as a source of knowledge for the system.

How the *inner-Object* artifact is built into *AGSys* is beyond the goals of this thesis. We just assume the characterization of *AGSys* in such a way that it is capable of exploiting knowledge models in order to build this abstraction. Nevertheless, this is a successfully studied field in the *ASys Project*, the wide framework within which this thesis has been developed (see Bermejo-Alonso [21], Bermejo-Alonso et al. [22, 23] among other works).

The relevant work of Bermejo-Alonso, includes the definition of all relevant characteristics—aimed to capture and exploit concepts—that might support the description of an autonomous system (a.k.a. *Ontology for Autonomous Systems (OASys)*). Even if *AGSys* is not necessarily considered as autonomous, some features of *OASys* are very useful in order to build—and currently to describe—the *inner-Object*.

As a general overview, *OASys* proposes two layers of abstraction in order to describe an autonomous system: (a) the *System Subontology*, regarding the problem of the *Body Architecture* and the knowledge related to it, and (b) the *ASys Subontology*, regarding the problem of the *Software Architecture* and the knowledge and processes related to it. And it is this *ASys Subontology* that contains the essential packages to specialize the *System Subontology* concepts, and that might implement the required packages in order to address the *inner-Object* problem. Usually, this is solved by using *top-level ontologies* or *foundation ontologies* which include all the relevant characteristics as descriptions of general concepts that are the same across all domains of knowledge [272].

Once built, the external object will become an existing object for *AGSys*, in the sense that this object becomes an *inner-Object*—in the form of a software artifact—that *AGSys* is able to exploit (see Fig. 5.14).

This inner-Object is just an artifact aimed to represent the external object. However, *AGSys* makes full use of this representation and derives actions from it. This inner-Object will be considered as the first form that an external object acquires during a perceptual process (as backed by several theories and discussed with regard to

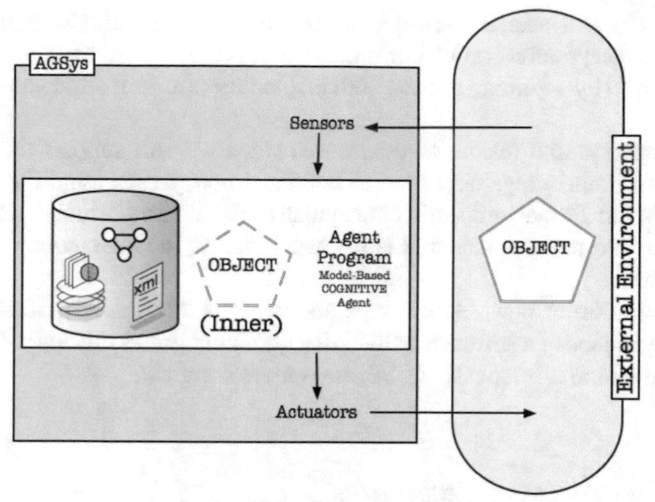

(a) inner–Object Representation

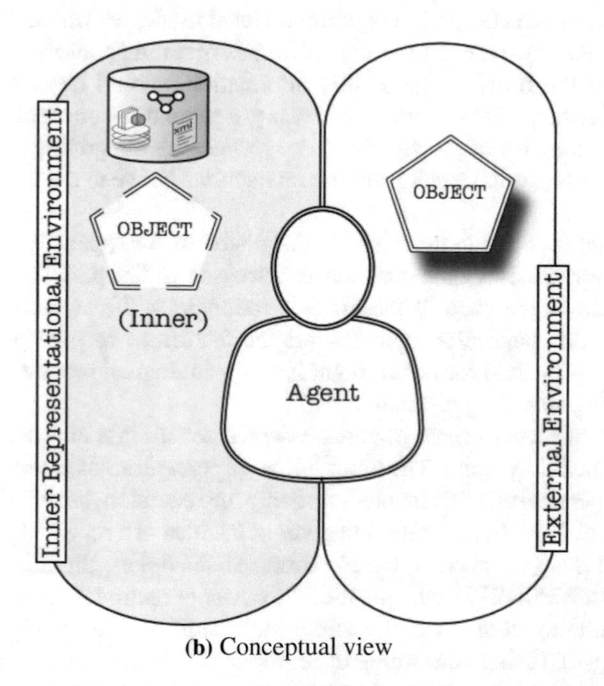

(b) Conceptual view

Fig. 5.14 Conceptualization of an inner-Object Representation: **a** From the point of view of *AGSys*, and **b** according to an easier conceptual representation

rationale ♣ 43). However, if this object is considered as the initial form, the question now is the conceptualization of the process that might act in order for this inner-Object to grow (by assuming growth both in quantity of information and abstraction degree).

This is an essential feature in this thesis, since we will support the viewpoint of Damasio Damasio regarding how to build *emotion*, *feeling*, and *the experience of feeling*. What Damasio does is conceptualize the recurrent integration of those relationships and patterns that are being created during the interaction of the object and the system.

The integration of new relationships and patterns into the inner-Object can be sustained by means of a software artifact that gradually grows this inner-Object, and by means of which it progressively becomes more complex.

5.5.2 *The Process of Perception*

The problem to solve is the design of computational processes which, fitted together, will integrate sensory–perceptual–conceptual information about the external object in *AGSys*. And the broad range of this information, should be wrapped into the *inner-Object* artifact. The reference to sensory–perceptual–conceptual information is analogous to the idea of growing the *inner-Object*, by integrating those new relationships and patterns that result from the interactions between the inner-Object and the system.

The essential foundation that justifies this vision of 'wrapped' is that perception is argued to be recursive. This question is discussed in Chap. 4, and in the debate about this topic in the state of the art (see rationale ♣ 30). Commonly, detailed theories about this particular topic assume the feature of recursiveness. And this thesis, regardless the real form that might have the biological process of perception, will accept this particular judgement.

The feature of recursiveness imposes essential constraints in order to design the analogy in artificial systems. The assumption of perception as a *recursive process* requires a deeper analysis that involves not only the recursion, but also how it affects the inner-Object. Initially, our intuition gives us the idea of a repeated process. Without a detailed analysis about different forms of recursion, there is wide general agreement with Wirth [273], who justifies the power of recursion in the possibility of defining "an infinite set of objects by a finite statement", an idea that is in accordance with the theory of Turing concerning the computation of continuous artifacts.

The debatable question herein is about the conceptualization of recursiveness, since even if the function seems to be characterized by recurrence or repetition, the argument is growing in complexity from repetition to repetition. The immediate consequence is that this conceived function of perception will apparently use different arguments at each stage (since the growth in abstraction changes the data type of the *inner-Object* as conceptualized in Fig. 5.15). Nevertheless, for the sake of an

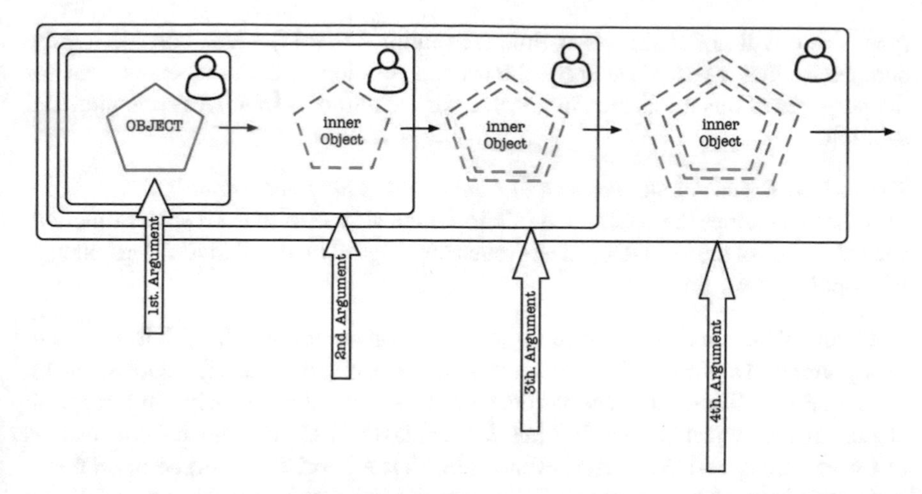

Fig. 5.15 Conceptualization about perception recursiveness and inner-Object growth (I)

easy explanation, for now we will conceive this function as a recursive function; but however highlightening this essential feature.

Since perception affects the form of the inner-Object (the artifact that allows for semantics regarding the real object), some type of interaction has to be assumed between the process of perception and this artifact. Apparently, if perception is generally assumed as a recursive function, this inner-Object can be conceived as an argument of this recursive function. We remark that this conceptualization of perception is applied in order to solve the problems that concern this thesis. There is no assertion herein about the real work of this complex process.

Continuing with our discussion, the argument for considering the inner-Object as an argument of a *perceptual recursive function* is allowing too for an additional feature. This conceptualization follows those principles of Turing regarding how to allow for a continuous inner-Object. And it directly derives in a possible way to build a continuous perception. But it imposes a new question about this argument.

Usually, computer science considers typed functions. That is, additionally to the body definition, functions are declared according to the values returned and the type of taken arguments. The problem of designing this recursive function of perception (the way it is conceived herein this thesis), is that our argument (i.e. the inner-Object) should continuously change concerning type from one to the following iteration in which it is used. The problem is not that inner-Object might change regarding its attributes, but that perception should involve structural changes that might change its type (see rationale ♣ 31 for theoretical foundations). And this, from a computational viewpoint, is a problem to solve.

We envision two possible solutions: (a) the integration, within the own body of the perceptual function, the action of changing both the inner-Object and the re-definition of the own function type, and (b) the conception of a topological type for the perceptual function, and the design of the inner-Object with this topological type which will not change over time (see rationale ♣ 31). Even considering the complexity that it states, we argue the second solution as the most convenient for the purposes of this thesis and, in general, for the purposes of a later computational solution.

Thesis 2 ♣ *Regarding the Perceptual Function and the inner-Object*
This thesis assumes the conception of a topological type of the perceptual function, and the design of the inner-Object with such topological type so that it will not change this topology over time.

Finally, what remains unsolved is the question about this topology. It is important to pay attention to the conclusions drawn in some perceptual theories that are related to this question. These theories consider some type of relationship between perception and the mathematical artifacts of *fractals* (see Hayek [105] and the final conclusions in López Paniagua [151]). And even more, those theories that argue a geometry of fractal shapes as descriptions of complex natural shapes (see March and Simon [159]) should be highlighted.

Without requirements of a deeper analysis for now, the referred topological type can be conceived as a set or a collection of elements [113]. And regarding this, there is a type of set that seems to be close to our requirements: the mathematical concept of *attractor set*. An *attractor* can be easily described as a collection of elements (i.e. a set) that attracts the results of an application over time as illustrated in Fig. 5.16 (see Boss [27]). With a general conceptualization, if an application f transforms some compact set X into itself, then the subsequent sets will be integrated one inside the one it follows:

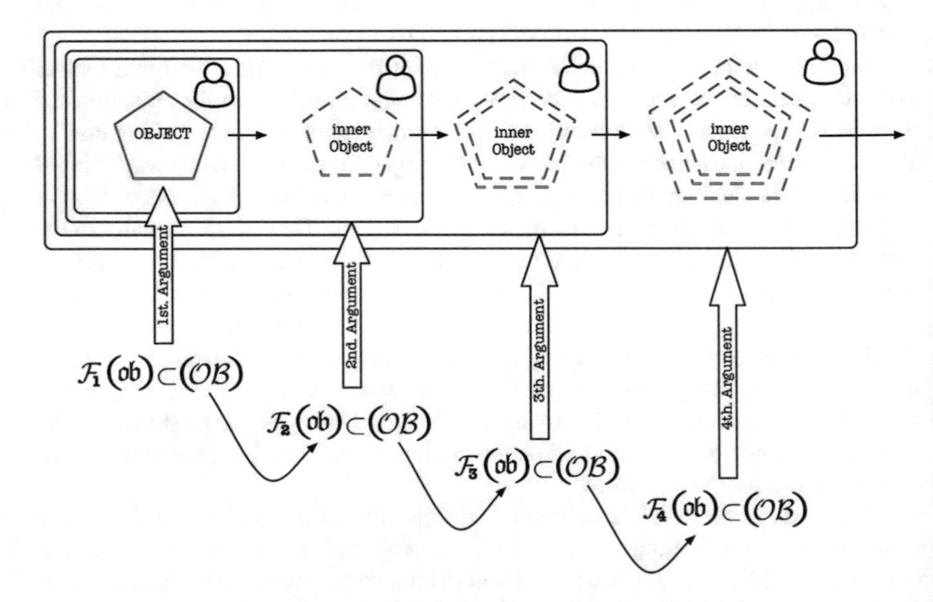

Fig. 5.16 Conceptualization about perception recursiveness and inner-Object growth (II)

$$\cdots \subset f^k(X) \subset \cdots \subset f(X) \subset X \tag{5.8}$$

It is important to remark that this solution is not argued as a definite solution, but as a proposed solution. The intuition of what it might be essentially requires a deeper analysis. However, it is relevant to note that it does not contradict those theories of perception previously referenced, since—even if it is not necessary—an *attractor* may be a *fractal* [27]. And it is also pertinent to point out that this idea provides a means to build a dynamic process of inner-Object growth, allowing for the characterization of this inner-Object as an invariant set. However, even assuming that this should be studied in depth, this conjecture serves for the purposes of our explanation.

Thesis 3 ♣ *Regarding the topology of the inner-Object*
The type of inner-Object is supposed to have the the form of an *attractor set*, in order to provide a means to build a dynamic process of inner-Object growth, and allowing for the characterization of this inner-Object as an invariant set. (Nevertheless, this possible solution is proposed for consideration and discussion)

5.5.3 Principles for Emotion Perception

We hold that—somehow—emotional assessment is an operation by means of which the quality of *value* is perceived. And 'quality' is herein assumed as an increased abstraction built from the source of a measurable difference. So the following question is how *AGSys* might build and integrate this *value*.

Along the analysis presented in the previous chapter, the amount of representational artifacts with which the emotional process in order to build emotion. As discussed in Chap. 4, there seems to be—however—a common agreement in the separation of the concepts of *emotion* and *feeling*. The way in which such assertion affects our work is that—according to such argued difference—*emotion* and *feeling* cannot be addressed from identical computational viewpoints, since they are referring to different domains and realities. By conceiving 'reality' as the space where a punctual abstraction is defined, we can argue that they are two realities of the same problem that require to be computationally integrated as different abstractions.

Consequently, the existence of two computational artifacts (without allusion for now to the form or type of these artifacts) will be assumed. In the description that Damasio presents regarding the process of *feeling perception*, he considers *emotion* as the previous artifact in order to obtain *feeling* [53, 54]. The process starts with the perceived object (in the multiples forms that he assumes for this object), to afterwards obtain the *emotion*. And later, the *feeling* construction is described on the basis of this *emotion*.

What is of relevance regarding this thesis, is a question that he asks considering the *emotion* as the new object regarding the *feeling*. Damasio conceives *emotion* as representations of the relationship between the organism and the [external] object.

And feeling an emotion results from “having mental images arising from the neural patterns which represent the changes in body and brain that make up an emotion” [54].

> The process that I am outlining is precisely the same we discussed for an external object, but it is difficult to envision when the object in question is an emotion, because emotion occurs within the organism, rather than outside of it.
>
> Damasio [54]

An argument that, additionally, is in accordance with the established logic of the perceptual process. Conceptually, analogous artifacts such as the *inner-Object* to build the *emotion-Object* and the *feeling-Object* will thus be conceived. The logic conceived for the perceptual process, allows for a description according to which *emotion* ‘serves’ to build *feeling* (see for explanations to Sect. ♣ 4.7.2).

I. Conceptualization of emotion-Object and feeling-Object Damasio distinguishes between *feeling* and *emotion* in two senses: (a) he argues that—likely from an evolutionary perspective, *emotion* appeared before than *feeling*, and (b) *feeling* performs a long-lasting effect in consciousness. And this phenomena of consciousness is argued to be responsible of integrating three stages of processes: (a) *“state of emotion”*, (b) *“state of feeling”*, and (c) *“state of feeling made conscious”*. This stage will be computationally described later.

Consequently, it is assumed that *inner-Object* should cause changes within the system. These changes should be represented later by the system, in a way that enables it to perform emotion and feeling (and later during the process, realizing consciousness of feeling).

Definition 5.7 ♣ *emotion-Object*
Computational artifact that we situate inside of the system, which accumulates the information about the emotion-based consequences that an external-Object is causing on the system.

Definition 5.8 ♣ *feeling-Object*
Computational artifact that we situate inside of the system, which represents an evolution of the *emotion-Object*.

II. From emotion-Object to feeling-Object Now, by assuming that the *emotion-Object* will serve to build the *feeling-Object*, and that they are two ‘realities’ of the same problem that require to be computationally integrated as different abstractions, a midway step that acts as an intermediate stage between these two ‘realities’ is required (see rationale ♣ 10).

Thesis 4 ♣ *Regarding the conceptual space between emotion and feeling*
The requirement of a midway step that acts as an intermediate stage between emotion-Object and feeling-Object is conceived. That is, a stage that keeps the result of the prime functional perspective in a form that can be used for the later functional perspective.

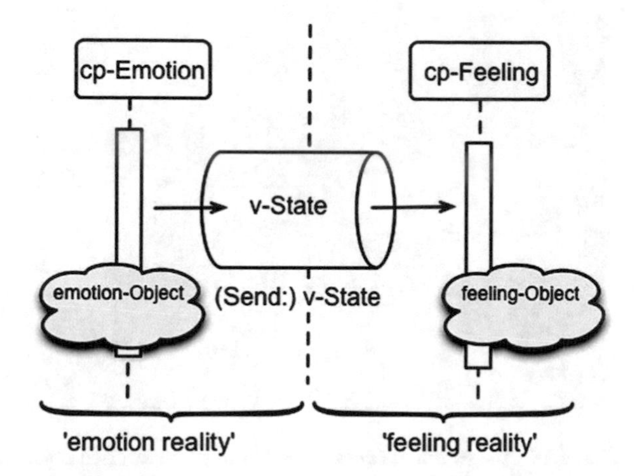

Fig. 5.17 Conceptualization of Computational Emotion (c-Emotion) and Computational Feeling (c-Feeling)

Definition 5.9 ♣ *valuable-State (v-State)*
Computational artifact which represents a midway step that acts as an intermediate stage between the *emotion-Object* and the *feeling-Object*. Computationally, this artifact might be described in the form of an abstract data type (ADT).

This *v-State* is conceived as a middle stage between the result of a process that builds the *emotion-Object*, and the starting point of the process that builds the *feeling-Object* (see Fig. 5.17). These two processes are *cp-Emotion* and *cp-Feeling* and they are conceived as:

Definition 5.10 ♣ *Computational Process of Emotion (cp-Emotion)*
Process by means of which the emotion-Object is built, under the argued rationale ♣ 8.

Definition 5.11 ♣ *Computational Process of Feeling (cp-Feeling)*
Process by means of which the emotion-Feeling is built, under the argued rationale ♣ 9.

5.5.4 *Principles and Baselines for Representation*

Firstly, we have to clarify the influence of an external object inside *AGSys* regarding the different conceptualizations that this object might take during the process of perception. Once more, we remark that this is a computational conceptualization made for the purposes of this work. This is not argued to be the real process that life systems perform to perceive, but just a way of understanding the assumed theories.

As Fig. 5.18 illustrates, the external object becomes a software artifact that *AGSys* is able to exploit by means of some *top-level ontology* that helps for its conceptual construction. This is the *inner-Object*.

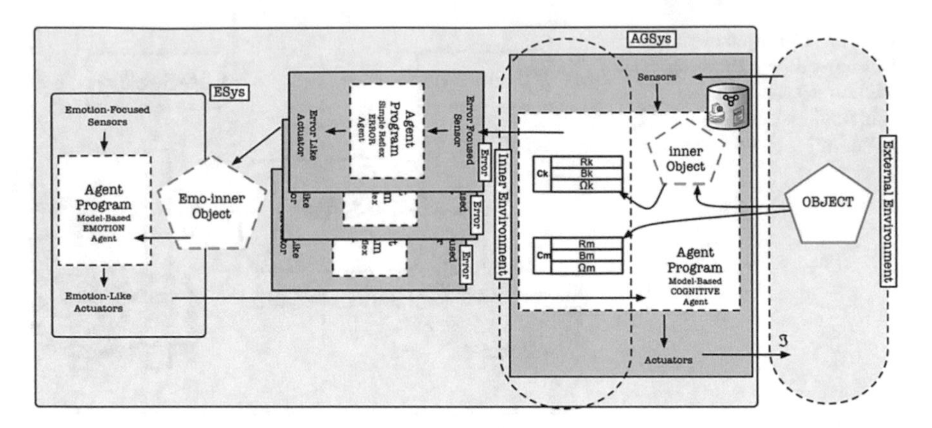

Fig. 5.18 The essential conceptualizations of an external object inside *AGSys*

This object, however, is causing two types of changes inside *AGSys* that are related to emotion: (a) changes directly caused because of the direct interaction of *ASys* with the external object, and (b) changes caused because of the recognition of this object in the form of the *inner-Object*. They both are interactions with *AGSys* that might cause emotional behavior, but this emotional behavior is assumed under different emotional realities. This will be described later in this chapter. For now, only the differentiation between these two realities is required (the partial model is illustrated in Fig. 5.19).

Therefore, what is analyzed regarding emotion are these two emotion-based effects on *AGSys*. That is, an *emo-inner-Object* that will be the starting point of the emotional evaluation.

Definition 5.12 ♣ *emo-inner-Object*
Computational artifact that we situate inside of the system, which represents those relevant changes that an external object causes on *AGSys* concerning emotion-based references.

Now, the following problem is the establishment of those interoceptive processes responsible of that subjectivity which Northoff [187] talks about. With the conceptualizations made in this thesis, the *emo-inner-Object* represents those changes that are relevant from an emotional point of view. This is a direct consequence of building the *inner-Environment* on the basis of emotional references.

Throughout the process of emotion perception, this *emo-inner-Object* will become: (a) an *emotion-Object* at a first stage (i.e. the *cp-Emotion*), and (b) a *feeling-Object* at the second stage (*cp-Feeling*) of the emotional perception process. These three objects are built on the basis of integrating the subsequent interactions of each artifact with *AGSys*.

By retrieving the description of the perceptual process, this interaction is conceived as a systematic work that recursively repeats specific processes. And these

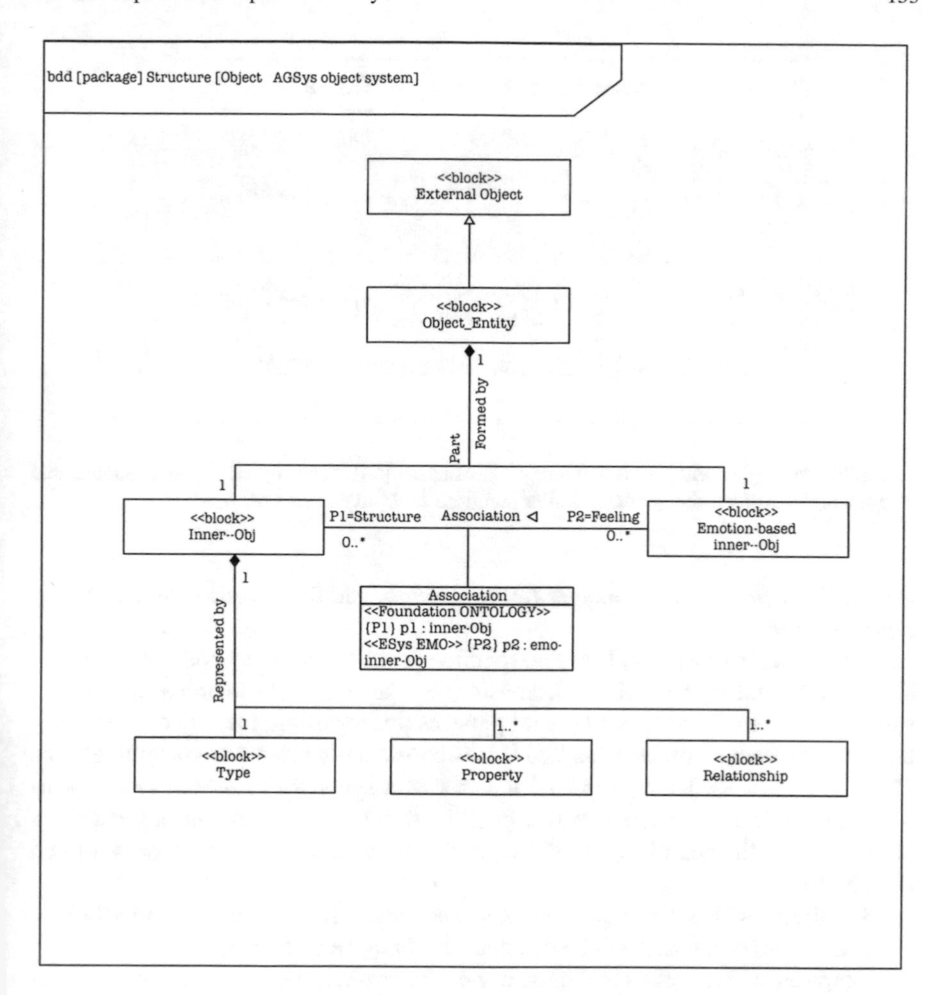

Fig. 5.19 Partial Model of an external object Structure inside *AGSys*

specific processes, will be featured by the abstraction degree of that stage at which each artifact is working.

There is a critical argument that affects some subsequent conceptualizations. It comes from the work of Damasio, which has been accepted in this thesis in order to comprehend the emotional phenomena: "There is no feeling without consciousness" [54]. So, we focus now on the idea of *imagery* because it is assumed within the referenced work as an abstract idea related to *thought* and *consciousness* (see 4.6.1 in Chap. 4).

The *emo-inner-Object* herein defined is a conceptualization made on the basis of the *bodily representation* idea of Damasio (see rationale ♣ 37). This is the representation of the bodily changes widely accepted in emotion science, and is postulated by the author as built within the work of two hierarchical levels of representation:

▼

ANALOGIES FOR NEURAL TERMS

CONCIOUS ⟺ Data susceptible of being obtained or used by *AGSys* in order to reasoning about it.
UNCONCIOUSS ⟺ Data exploited by *AGSys* in order to attend processes but NOT valid for reasoning about it.

MAPS ⟺ Neural Patterns (Like) ⟺ Relationships perceived from outside of *AGSys*
IMAGES ⟺ Mental Patterns (Like) ⟺ Relationships perceived by *AGSys*

IMAGES can be CONSCIOUS or UNCONSCIOUS

▲

Fig. 5.20 Analogies with the terminology of Damasio [54], focusing on defining comparable computational artifacts that perform similar functions in artificial systems

(a) The *First-order bodily changes representation*, and (b) the *second-order bodily representation.*

As previously described, Damasio tried to show how our consciousness works to allow for mental *representation*. Damasio uses the term *representation* making no suggestions about how faithful neural patterns are regarding the objects they refer to, and defines the term as a "pattern that is consistently related to something" (see Chap. 4, Sect. 4.7). He uses this term either as a synonym of *mental image* or as a synonym of *neural pattern*, even though he differentiates that *neural pattern* (or *map*) refers to the neural aspect of the process by which a *mental image* is formed (Fig. 5.20).

This thesis will use analogies of *map* and *image* referred to by Damasio (see rationale ♣ 38). And they will be denoted as *c-Image* and *c-Map*.

Computer science refers to the use of *patterns* from several points of view. However, *patterns* generally allude to:

1. *Patterns of Relationships* Relationships among elements (either data, cues, systems, subsystems, …).
2. *Patterns of States* States concerning sets of elements.

We will also differentiate between those processes that sustain the system from those that define the system knowledge. As in biology, we will differentiate conscious and unconscious processes by establishing the difference between (a) system-knowledge and (b) system-sustaining processes. The former are driven to make inferences by the system, and the latter are driven to sustain the operation of the system. This is in accordance with Turing, who assumed that machines are conscious only about what they can *'read'* [259]. Herein, this idea has been however extended to *'knowledge'* (see rationale ♣ 47).

Thesis 5 ♣ *Conscious vs. Unconscious*

1. By accepting the view of Damasio, we will assume that some of these constructs conceptualized as *(c-Images)* will be directly accessible by the system in order to be used as knowledge. This is the computational conceptualization of *'being conscious of'* or *'being capable of use'* the information as knowledge.[2]
2. Consequently, we will assume—in reverse way—that some of these constructs conceptualized as *(c-Images)* will not be directly accessible by the system. This is the computational conceptualization of *'not being conscious of'* or *'not being capable of use'* the information as knowledge.
3. When Damasio refers to 'accessibility', he argues that "conscious images can be accessed only in a first-person perspective (my images, your images)", while "neural patterns, on the other hand, can be accessed only in a third-person perspective". Analogously, we will assume that accessible *c-Images* or *conscious c-Images* will be retrieved only by the system and for the use of the system, in terms of *'used as knowledge base in order to make inferences'*.[3]
4. Even when *c-Maps* and *c-Images* are conceived as different artifacts (relationships and states respectively), we argue that they both are used [by the system] in a similar way. They both are used [by the system] to provide mechanisms in the operation of assessment. The inner processes of evaluation use *c-Maps* and *c-Images* to infer results that are used afterwards by the system. According to the idea of Damasio about the lack of consciousness in these data, the computational system analogously will not use *c-Maps* or *unconscious c-Images* in order to make inferences.

Definition 5.13 ♣ *artificial consciousness*
Any type of information susceptible of being exploited by any rational system such as *AGSys*

Hereafter, whenever this work employs the term *consciousness* regarding *AGSys*, it should be assumed that it must be interpreted as *artificial consciousness*

Definition 5.14 ♣ *computational-Map (c-Map)*
Patterns of relationships concerning a set of system artifacts. They are patterns that (a) are not conscious for *AGSys*, and (b) serve to build *c-Images*.

Definition 5.15 ♣ *computational-Image (c-Image)*
Patterns of states concerning a set of system artifacts. They are featured to be either conscious and unconscious.

---

[2]We remark that there are hard differences between this conceptualization and the one that refers to the *self-consciousness*.

[3]We are referring herein to inferences that involve methods at the intersection of machine learning, artificial intelligence, bayesian inference, probability, statistics or knowledge base systems among others.

5.5.5 *Summarized Problem*

Emotion is not a question of *'knowing value'* but a question of *'perceiving value'*. This thesis assumes the ability to obtain value as an ability closely related to the *perception* of this value (this is a thesis argument ♣ 1).

This section can be summarized taking into account three main categories of problems that need to be solved to approach a process which allows *AGSys* for emotion perception:

1. The *inner-Object* and *emo-inner-Object* as the prime symbolic artifacts.
2. The problem of a function that approaches *perception*.
3. The problem of correlating neural versus computational terms.

The first problem is the construction of the prime symbolic artifact which *AGSys* will use as knowledge: the *inner-Object*. This artifact is associated to the *emo-inner-Object*, which integrates knowledge about the first stage of effects that the external object causes on *AGSys*.

The second problem is the computational approach to perception in order to build *feeling*. This has been addressed by considering a recursive function typed as *attractor set*.

The last problem is the correlation of those neural terms that come from the emotional theory of Damasio. This theory has been accepted as the support of the proposed model. Therefore, this correlation is an essential problem in order to design the main principles of the theory. The following section will use the rationale herein described.

5.6 The Inner Agent ESys

As introduced in Sect. 5.3 of this chapter, the conceptualization of transversality together with the intuition that emotion works inside systems [227, 231], has resulted in the solution based on an inner agent *ESys*. This agent is conceived as embodied into—and living inside—*AGSys*, and aimed to perform the complex transversal work of emotion assessment.

The generic form of *ESys* is that commonly assumed to be presented by rational agents. As *ESys* is hosted in *AGSys*, the environment of *ESys* is the inner 'reality' of *AGSys*. Therefore—regarding the normal work of a rational agent, the *utility*, *performance measurement*, and *utility function* are related to those sequences of the inner situational states of *AGSys* (rather than those sequences related to the external world). Essentially, the work of this inner agent will aim the monitoring of those values that the *Error-Reflex agents* provide, and change these values from its current form of error, to another form that allows for emotion-based assessment.

Emotion assessment forms a complex system of different processes as necessary parts in order to build the emotional process, but focusing on the major goal of artificial emotion experience. Even realizing the real limits of *experience* in artificial systems from the very starting point of this explanation, this thesis assumes the work of *ESys* aimed to provide the whole set of elements that might be required by *AGSys* in order to build this *experience*. Therefore:

Thesis 6 ♣ *Regarding the Role of ESys*
The role of the inner agent *ESys* focuses on providing *AGSys* with the essential components in order to build the emotional experience. *AGSys*, however, is considered as the responsible of the *'emotion observation'* or *'emotion perception'* (under principles of optimal operation). The role of *ESys* is bounded to provide the required elements for this observation.

I. Assessment versus Inference As previously argued, this thesis assumes a *cognitive approach of emotion* (see rationale ♣ 3). It accepts that *AGSys* is able to assess events that come from the external environment on the basis of its own concerns.

In this approach, the conceptualization of *appraisal* will be assumed as an essential feature of the emotion experience (see rationale ♣11 and ♣ 11). And the main features that characterize this *appraisal* will be defined on the basis of the set of reasons that result from the analysis of emotion in Chap. 4.

As previously introduced in this chapter, this work will employ the term *artificial emotion* to refer to the complete phenomena realized in life systems (i.e. in the sense of 'a system that deploys artificial emotion'). And it will be differentiated from *computational emotion* in the sense that the former will refer to the complete phenomena that an artificial system might deploy, and the latter will just refer to the entire computational work that sustains the former (see Fig. 5.7).

II. General Assumptions It will be assumed that *emotional assessing* will not be made within the same functional structure as that related to the *inferential processes* (see rationale ♣ 15). They will be considered as two different 'realities' since the former reasons regarding emotional references, and the latter should reason regarding the former reasoning results.

One common argued feature of emotional behavior is that several responses of a concrete agent can vary depending on the situational state of this agent and its environment. This results in different emotional responses under the same conceptual emotion.

If *emotional assessing* and *inferential processes* are integrated within the same structure, a *mapping-function* seems to be required in order to allow for these variations. And it might be computationally unviable. Computational systems optimize their work within structured systems that are intended to do a specific work. The inference on the basis of an emotional assessment is thus better, rather than the mapping of all possible responses.

Thesis 7 ♣ *Assessments vs. Inferences*
The role of the inner agent *ESys* focuses on providing *AGSys* with the essential components in order to build the emotional experience. It is assumed that *AGSys* will reason regarding the assessments made by *ESys*. And *ESys* will reason regarding the defined emotional references.

5.6.1 *The Appraisal*

Appraisals will then be assumed to be built on the basis of these two assessments: (a) the assessment regarding the situational state of the *emotional references*, and (b) the assessment regarding the situational state of *AGSys* (that is, how these assessments are affecting *AGSys*). Therefore, it will be built on the basis of these two essential evaluations performed by *AGSys* and *ESys*.

Thesis 8 ♣ Regarding the Appraisal assessments
It is assumed that *Appraisal* is built on the basis of the work performed by *ESys* and *AGSys*, as a whole evaluation of both the emotional references and the system references.

I. Time as a variable Time is assumed by theorists as an essential variable that influences the affective states, by considering the time during which they occur. Even if out of the scope of this thesis, time is a distinguishable factor of an emotional event [96]. It is an attribute that helps make high order emotional distinctions between *moods*, *temperaments*, *attitudes* and *emotions*. And it also influences the normal work of the emotional process, by causing effects on several factors such as the *copying* or—to give an additional example, processes of re-appraisal that might result in new emotions and that might influence *motivation* (see rationale ♣ 20).

We thus subscribe to the concept of time as an essential feature of emotion in order to allow for the emergence of some appraisal dimensions such as *copying*, as well as re-appraisals that might result in new emotions and that might help control the state of *motivation* of the system.

Emotion intensity is also featured by the speed of the change, since the spectrum of the affective situations that may be happening are related to the duration [190]. Even more: Depending on the speed of effects, the same event coming from the environment is assumed as the cause of different emotional responses. That is for example the increment regarding a noise in the environment. If the noise grows slowly over time to a concrete annoying intensity, the emotional response will evolve slowly towards states of annoyance, irritation, exasperation or so. However, if the same final intensity is raised at a given time without expectance of that noise, the emotional response will be close to fright or shock.

Therefore, the conceptualization of the whole model herein described with such a characterization should take into account *time*. It is an essential element of the emotional assessment nature. An emotional assessment cannot be defined without the influence of time, since assessment itself as well as later processes derived from it are intrinsically affected by this concept of '*time*' (see rationale ♣ 20).

This integration of time, is somehow understood as the integration of dynamics as a property that stimulates *AGSys* toward a specific emotional response. This refers not to system-based dynamics, but to emotion-like dynamics.

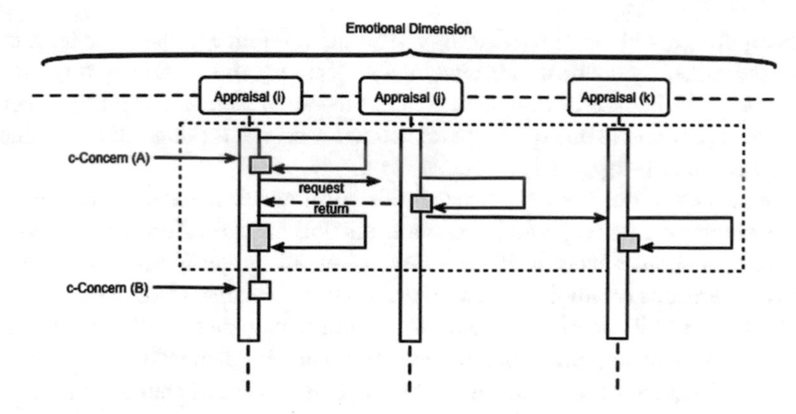

Fig. 5.21 Two concerns cannot be concurrently appraised. However, there can be several parallel appraisals for the same concern

Thesis 9 ♣ Regarding Time
Time is considered as an essential feature of appraisal. It will be incorporated by the integration of dynamics as a property that stimulates *AGSys* to a specific emotional response.

The question of time does not refer to the question of letting the system an interval of time by design to resolve challenges. It is a question that regards the influence of this time on the emotional assessment as an additional variable (see rationale ♣ 24).

II. Concurrency The work of appraisal will be accepted from a mixed perspective in which: (a) processes of slow and fast appraisal are concurrently occurring, and (b) these processes are not discrete but they are considered as different appraisals working at the same time. However—according to Ellsworth and Scherer [71]—no assessment of two *c-Concerns* at the same time will be assumed. By contrast, several appraisals working concurrently in parallel for a concrete *c-Concern* can be considered (see Fig. 5.21). They will take different computational times depending on the resources that they need in order to work (see rationale ♣ 17).

III. Dimensions, concerns and emotional goals Until now, several elements of our reference model have been described on the basis of an artifact which we have referred to as *'emotional references'* or *'emotional goals'*. However, one essential aspect related to our approach is the universe of characteristics depeding on which an artificial emotion can be assessed (see rationales ♣ 22 and 14).

Thesis 10 ♣ Universe of Emotion
A universe of emotion where all characteristics of emotion assessment are considered as a whole is assumed. The coarse grain characteristics are assumed to be the references used for the assessment.

Therefore, the following paragraph will describe how these references for emotion are defined. Emotion science describes emotion appraisal from several perspectives and using different kinds of terminology. And the reference to the same term is not always the same. By contrast, it is absolutely necessary and extremely important to build an appraisal space of evaluation: the *Universe of Emotion*. And the essential aspects in order to build this space are related to the search for those pillars by means of which emotion is triggered and categorized.

As described in the previous chapter, Ellsworth and Scherer conceive appraisal as performing an essential role in emotion elicitation and differentiation, with no clear position about which the central dimensions of appraisal are. Only the categorized dimensions of *novelty*, *valence*, *goal/needs*, *agency* and *norms/values*—are described, as a small part of those major dimensions postulated by different theorists. Clearly, these concepts pertain to different realities and implement differences in their nature (i.e. abstraction). Consequently, this explanation establishes a computational problem that requires to be solved.

By paying attention to the argument of Ellsworth and Scherer, concurrency is not allowed regarding concerns. However, for a single concern evaluation, parallelism is allowed. Under a common understanding, *novelty*, *valence*, *goal/needs*, …seem to represent aspects and features regarding some environmental situation. That is, a concrete environment enables a series of circumstances that might affect the emotional state of *AGSys* such as: the 'novelty' of a concrete noise or the 'need' of being supplied.

The concept of *dimension* is argued to be a measurable extent of emotional assessments. That is, a mode of extension that is present in several ways in the emotional space (i.e. *novelty*, *valence*, *goal/needs*, …). The illustration of this idea is described in Fig. 5.22.

The term *concern* refers to something that is relevant. Somehow, it serves to build the metric in order to measure a concrete *dimension*. And *goal* makes reference to some type of *'score'* that can be applied to measure the accomplishment of something. This concept might thus help to build the metric in order to measure a concrete *concern*.

This is a categorization that serves for the purposes of this thesis. However, this is not the universal semantic usually—or commonly—applied to *dimension*, *concern*, and *goal* regarding emotion appraisal. Therefore, we will use the namespace label of *'computational'* in order to establish proper semantics for this current work.

By designing a conceptual algebraic model for representing the emotional dimensions like vectors, these vectors will constitute the identifiers of the situational state of emotion. This concept of 'algebraic model' is however related to a mental concept used herein to describe an idea. And this idea regards the use of emotional dimensions in order to build an artificial concept of emotion. It, however, can be conceived within any other type of space that might provide analogous universes.

Definition 5.16 ♣ *Appraisal Dimension (a-Dimension)*
Analogous to the concept of vector in a vector space, an appraisal space formed by a collection of N *appraisal-Dimensions (a-Dimensions)* is assumed. This will define the number of independent directions of emotional assessment for *AGSys*.

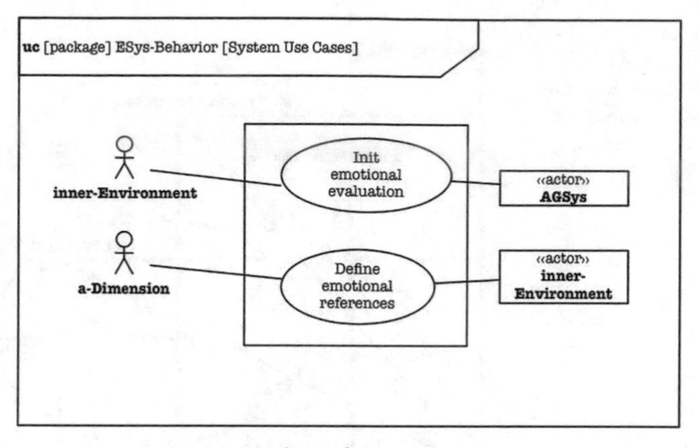

(a) A sample use case

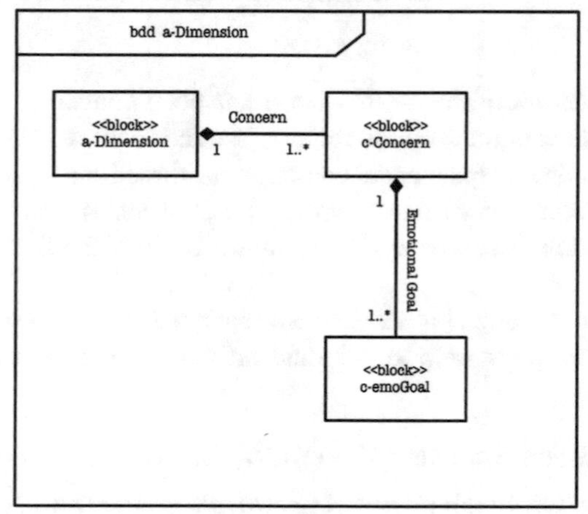

(b) a–Dimension actor

Fig. 5.22 Conceptualization of a-Dimension, c-Concern and c-emoObject

Definition 5.17 *Computational Concern (c-Concern)*
It refers to a collection of functional artifacts used to build a metric with which to measure a concrete *a-Dimension*. This measure is conceived in terms of positive or negative distance to those matters of importance for that *a-Dimension*.

Finally, *c-emoGoal* is conceived as some type of *'score'* that can be applied to measure the accomplishment of something. The *Error-Reflex agent* is proposed as the artifact that will measure the *inner-Environment* of *AGSys*. Therefore, what represents these *c-emoGoals* is the function based on which some *appraised subsystem* $S'_{appraised} \subseteq S_{AGSys}$ is scored (see Eq. 5.5)

Definition 5.18 *Computational emotion-like Goal (c-emoGoal)*
The value that uses the system to measure the positive or negative error regarding some c-Concern.

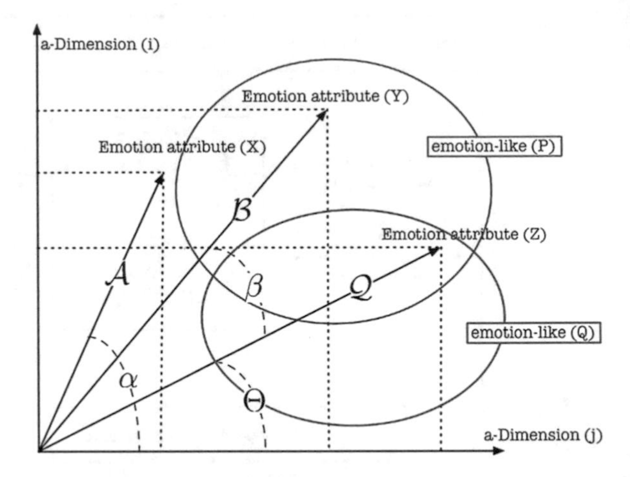

Fig. 5.23 A single conceptualization of a generic *Universe of Emotion* described under dimensions of assessment

IV. Universe of Emotion The *a-Dimension* has been conceived as a collection of *appraisal-dimensions* in the form of vectors (see rationale ♣ 21). The justification is to define the number of independent assessment-directions in the appraisal space, i.e. the appraisal vector of attributes within the *Universe of Emotion*. For the sake of a clear explanation, it will be explained on the basis of the illustration showed in Fig. 5.23.

By assuming *m* directions of assessment, each *a-Dimension* represents a separate attribute which might help to compute artificial emotion as a composition of attributes:

$$\begin{aligned}&\langle\text{Emotion Attribute}\rangle(X) = (aD_1, \ldots, aD_m)\\&\langle\text{emotion-Like}\rangle(P) = \amalg_i(\langle Emotion - Attribute\rangle(i))\\&\text{being i defined as: } i = (X, \ldots, Y, Z, \ldots)\end{aligned} \tag{5.9}$$

This approach serves to compare emotions with dimensions. This way, we can model how appropriate any *a-Dimension* is to the matter of a specific emotion (see Fig. 5.23).[4] In order to provide an example, we will assume an emotional theory within which to conceptualize:

$$\begin{aligned}&\langle\text{Emotion Attribute}\rangle(X)\\&\langle\text{Emotion Attribute}\rangle(Y)\end{aligned} \tag{5.10}$$

This example assumes a previous operation along which the values of these emotional attributes have been computed by the system. Therefore, their components can be obtained and expressed in the following form:

[4]The requirement for a theory of emotion in order to establish the assumptions that regard artificial emotion concerning *a-Dimensions* is strongly highlighted.

$$\begin{aligned} \langle \text{ Emotion Attribute}\rangle(X) &= (aD_1, \ldots, aD_m) \\ \langle \text{Emotion Attribute}\rangle(Y) &= (aD_1, \ldots, aD_m) \end{aligned} \tag{5.11}$$

Consequently, emotions P and Q might be modelled under the support of this same emotional theory as follows:

$$\begin{aligned} \langle \text{emotion} - \text{like}\rangle(P) &= \mathcal{F}(\cos(2\theta), \mathcal{A}, \alpha, \ldots) \\ \langle \text{emotion} - \text{like}\rangle(Q) &= \mathcal{F}(\sin(\theta), \beta, \mathcal{B}, \ldots) \end{aligned} \tag{5.12}$$

In order to construct the *Universe of Emotion*, it is required to incorporate the artifacts of *Emotion Attribute* and *a-Dimension* in such a way that enables their computational use. That is, they should have the appropriate form to enable models on the basis of multiple theories. This problem regards two essential issues: (a) what the *Emotion Attributes* are and how they are defined, and (b) what the *a-Dimensions* are and how they are defined.

1. *The a-Dimension*
   It is conceived as the frame within which *c-Concerns* and *c-emoGoals* are defined. The *Universe of Emotion* is conceived in this thesis as a model based on linear algebra. However, each *a-Dimension* needs to be built on the basis of *c-Concerns* and *c-emoGoals*, and it is visualized by using Model-based tools. This will be described later in this chapter, once the conceptualizations of *c-Concerns* and *c-emoGoals* has been shown.
2. *The Emotion Attributes*
   They are conceived as the basis of an emotion 'quality', in the sense that they represent the state of a concrete inherent part of the emotion assessment regarding each emotional dimension (i.e. *a-Dimension*). *Emotion Attributes* are conceived as universal features concerning the whole set of *a-Dimensions*, and this requirement is related to the metric used to measure within the *Universe of Emotion*. Assuming (m) *a-Dimensions*, each *Emotion Attribute* is formed by a number (m) of components. Thus, in order to maintain coherence regarding what this attribute represents, it is needed that each component represents the same characteristic related to each dimension.

To give an example: Since any *a-Dimension* is defined by means of *c-Concerns* and *c-emoGoals*, a concrete *Emotion Attribute* might be the 'positive value of goals'. This way, within the framework of a concrete *a-Dimension*, processes will be computed in order to obtain a characterization of the state of this attribute. This way, the component of the *Emotion Attribute* in this concrete *a-Dimension* is provided.

Some additional examples of *Emotion Attribute* might be: *c-emoGoals* stability, *c-Concerns* stability, estimated evolution of these stabilities, ...

V. Summary of appraisal features *Appraisal* has been approached in the form of a space model named the *Universe of Emotions* within which *Emotion Attributes* can be defined in order to transversally characterize common features that affect all emotional dimensions. That is, emotion is commonly affected by these common features regardless dimensions: the accomplishment of regarded goals is required in all emotional dimensions. They are different goals, but the accomplishment is always required.

The use of *Emotion Attributes* allows for the advantage of integrating weights in order to conceptualize each final emotion, but this in not the only advantage. Once the *Emotion Attributes* are defined, they can be assessed according to additional variables such as angles, modules, etc. in order to obtain several ways of characterization.

This conception of a *Universe of Emotions* allows for areas within which emotions are not exactly determined, enabling this way for an 'undetermined' number of emotional states without label but with category. And this is in accordance with rationale ♣ 19 that regards the requirement of a continuous appraisal and emotionality.

The approach herein described allows for: (a) the assessment as the essential core of emotion triggering but not the only cause, (b) the idea of appraisals as components of emotion, since those dimensions in which the assessments are made constitute the essential core of the emotion categorization, (c) the continuous assessment according to the idea of a continuous character of emotion, (d) an agreement with the existence of emotionality, that is, feelings that are not filling the pattern of a concrete category of emotion, and (e) time as an essential appraisal's feature.

This proposal is in accordance with the following essential attributes of the emotional assessment:

1. The emotion will be generally accepted as a multi-component system whose parts are intended to achieve functions of different nature: physiological, cognitive, qualitative and subjective responses (see rationale ♣ 2).
2. A continuous process of assessment will also be assumed. Emotion is considered as a direct consequence of *appraisal* (see 5).
3. The idea of a multi-dimensional process of assessment will be assumed (see rationale ♣ 5).
4. The idea of a multi-dimensional intensity of assessment will be assumed (see [82].
5. Dynamics associated to emotion will be assumed. These dynamics will not be conceived as *AGSys* dynamics, but as the dynamics that the situational state of emotional goals provide to *AGSys* (see rationale ♣ 20).

5.6.2 The Emotional Goals

Artificial systems work under principles and foundations different from those that characterize life systems. One essential advantage of artificial systems is that they are 'designed systems' (an apparently nonsense assertion that actually is not such). The emotion operation is aimed to organize and manage system priorities, to maintain system wellness, to make full use of the system capabilities (and derive benefits from its operation), or to monitor and assess its own actions, among several other focuses of activity. The real phenomenon of emotion, sustains different levels of intensity and time periods that allow for an intricate source of grounds for behaving in a particular way.

Emotional goals seem to be an essential piece of this transversal operation. However, since it is not clear how the brain performs this transversality, the solution might be to convert the goals into rational agents. Regarding the fact that artificial systems are designed, the advantage is that it allows engineers to implement emotional goals as "emotional workers". And let they perform some specialized tasks of emotion phenomena in a transversal way during the normal operation of the system.

Emotions are characterized by the widest range of dynamics provided to the systems, and it seems as if the very central cause of these dynamics were related to the emotional goals. These dynamics are usually associated with the specified item of goal accomplishment. Sometimes, these dynamics are supposed to have the amount of physiological mechanisms concerned with the motion of the bodies as a result of an emotional response. A type of mechanism that justifies a lot of theories that regard the embodiment of emotion [107], and which is widely aimed to be modeled. However, this is not the only kind of dynamics affecting a system during an emotional response.

Analogously to the idea of Schelling [232], our conceptualization of emotion emerges from the idea of a system of goals that help to endow *AGSys* with motives for movement (i.e. for triggering actions), and the assessment of the macro behavior to assess equilibrium as the ground for an emotional characterization.

Models should enable systems for those dynamics associated to the action related to the issue that causes the emotional behavior. It is not the association between the cause and the action, but the association between the cause and the dynamics towards the action. They are two options that endow *AGSys* with sematics. This results in emotional-like dynamics associated to the emotion itself, as an effect of the system's designs. And from the viewpoint of this thesis, this effect seems to derive from the behavior of the emotional goals.

By now, a mirror model of the brain work is hard to be designed. What can be done, however, is transfer this operation of dynamics to the goals themselves, and this way let *AGSys* analyze the result of these dynamics. With this conceptualization, it is easier to analyze the dynamics of goals that are not in equilibrium, understanding relationships between individual states of goals, interactions among goals, etc.

I. The system of emotional goals From our engineering perspective of *Model-based systems engineering (MBSE)* we will conceive a generic form for goals in accordance to the vision of an *Agent-Based Model*.

Generally speaking, computer science conceives *agents* as computer programs that interact with other agents (either artificial or live agents) in order to produce particular results. In agreement with the view of Downey [65], what specifically characterizes an *agent-based model* is that: (a) the agent models intelligent behavior by using sets of rules, (b) this agent interacts with other agents, (c) works by using incomplete local information, (d) models use to include variability between agents as well as (e) random elements. Downey argues however that they are useful for (a)

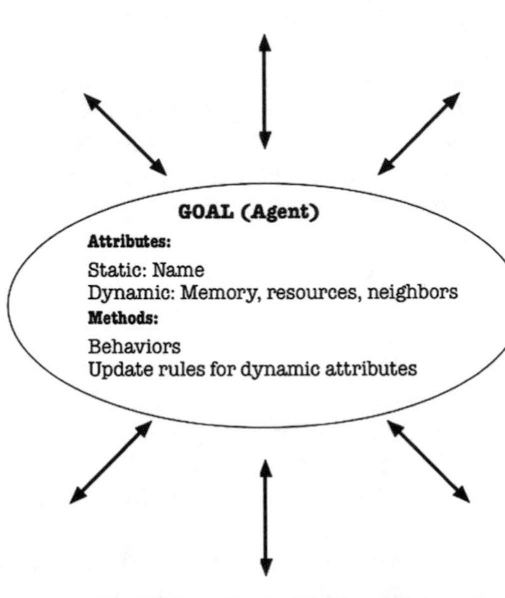

Fig. 5.24 A single view of an agent based goal (*Source* under the common understanding of typical agent based on Macal and North [154])

modeling "the dynamics of systems that are not in equilibrium (although they are also used to study equilibrium", and (b) for "understanding relationships between individual decisions and system behavior" [65].

Thesis 11 ♣ *Goals as Agents*
Goals are assumed as agent-based models in order to produce results related to the dynamics of an emotional response. The center of the emotional assessment is transferred from the system to the emotional goals.

Therefore, *c-emoGoals*, *c-Concerns* and *a-Dimensions* will be conceived in the form of agent based models, in order to provide *AGSys* with some essential abilities of rationale regarding the situational state of the emotional goals (see Fig. 5.25).

II. The model of emotional goals As the structure of the system has been conceived, each type of agent (i.e. *c-emoGoal*, *c-Concern* and *a-Dimension* types) should interact with its own environment with agents of the same type.

As it is explained in Macal and North [154], these three environments will be used to provide information of spatial location of an agent in order to evaluate relative positions between agents of the same type, positions regarding the environment, etc (Fig. 5.24). This allows for the design of frameworks—in the form of environments—within which the information about, and the interaction with agents may be improved.

The distinctive feature of our particular approach is a hierarchical structure where *c-Concerns* are built on the basis of *c-emoGoals*, and—at the same time—the entire series of *c-Concerns* are representing an emotional dimension *a-Dimension*. This

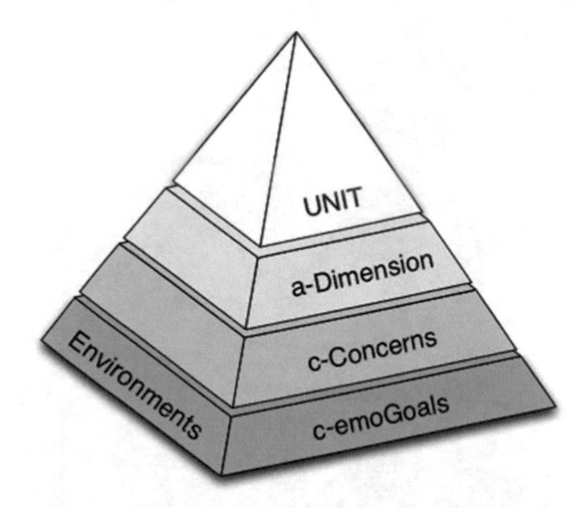

Fig. 5.25 A single view of the agent based structure of an emotional goal

represents a hierarchical structure of goals and environments that might provide essential information for the construction of the *Emotion Attributes* related to the *Universe of Emotion*.

This way, we propose a three layer infrastructure of goals that improves the work of *appraisal* if emotion is referred to, and the possibilities of model-driven if design is referred to. In this manner, the work of each part of the emotional assessment is partitioned in such a way that it enables complex behaviors.

The essential operation of the entire framework is conceived as follows: (a) each individual environment related to a type of goal will provide the framework for that type of goal, (b) each layer is cohesive and depends only on the layers below, and (c) the cohesiveness in each layer enables agents interaction either among them or within the geographical position in the environment. A general portrait of this framework is showed in Fig. 5.25.

This way, according to this description, pyramidal frameworks related to each *a-Dimention* with a behavior characterized by the way in which *c-Concerns* and *c-emoGoals* conduct themselves can be obtained. With this supporting structure of the emotional goals, different characterizations of states can be computed in order to obtain values to build vectors of *Emotion Attributes*.

As previously described, a concrete *Emotion Attribute* represents a vector of components of a specific feature. Thus, this vector represents a transversal viewpoint of this feature regarding all *a-Dimensions*. Hence, in order to maintain a coherent meaning for each *Emotion Attribute*, the feature requires to be measurable within the frameworks of any *a-Dimension* considered. As previously exemplified, these features might be *c-emoGoals* stability, *c-Concerns* stability, estimated evolution of these stabilities, …(see Fig. 5.26).

Thesis 12 ♣ *The Structure of Emotional Goals*
The emotional goals are conceived according to an agent-based model with pyramidal structure. This way, the emotional state related to a concrete dimension of appraisal

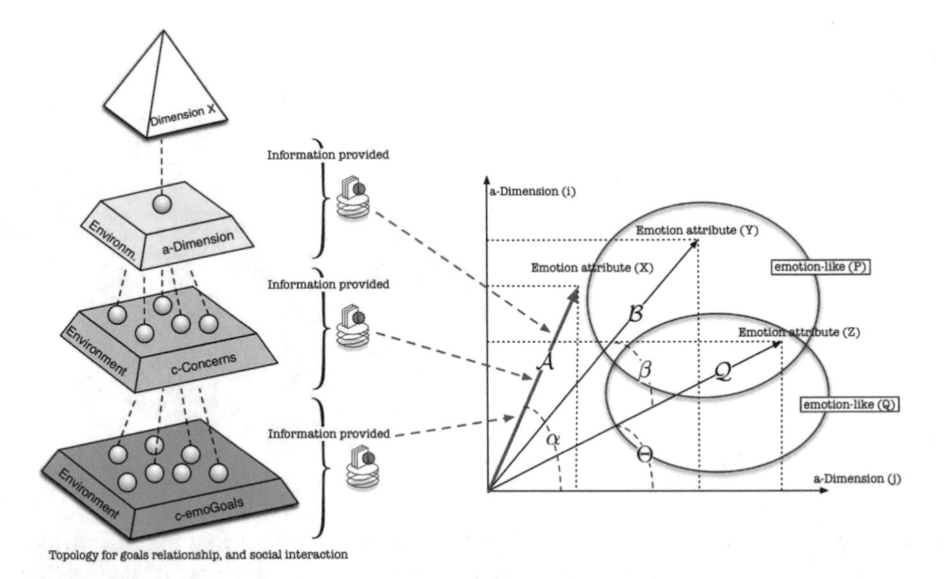

Fig. 5.26 An extended view of the agent based structure of an emotional goal and the relationship with the *Emotion Attribute* vectors

is the collective state of all agents (i.e. *c-emoGoals*, *c-Concerns* and *a-Dimensions*) along with the state of their environments.

III. Components of the system The aim is to model the dynamical system of interacting goals, so each individual artifact should perform its own actions following the discrete events caused by errors in goals accomplishment. The function of 'what should be accomplished' has been integrated as part of the *Error-Reflex agent*, and this agent measures the *inner Environment* of *AGSys*. Even if this *Error-Reflex agent* is not part of the framework of emotional goals approaches, this agent is the one that provides this discrete event of each *c-emoGoal* as illustrated in Fig. 5.27. Herein we remark once more that *c-Concerns* are built on the basis of *c-emoGoals* behavior.

We conceive a system of emotional goals whose equilibrium determines the emotional state of *AGSys*. Those principles that help build a *Universe of Emotion* within which to characterize the final emotion have been introduced. However, it is not enough in order to obtain the *'feeling'* and the *'feeling of feeling'* to which Damasio refers.

As a coarse grain summary, the emotional response is usually defined by essential responses that regard dynamics, triggering of actions related to the situational state, perception of goals accomplishment, emergence of new goals related to actions, memories interaction or resources management for action readiness among several others.

We defend designs that are close to the biological principle of optimization. That is, topologies of work susceptible to be replicated whatever the level of abstraction. Consequently, we consider that *c-emoGoals*, *c-Concerns* and *a-Dimensions* will all

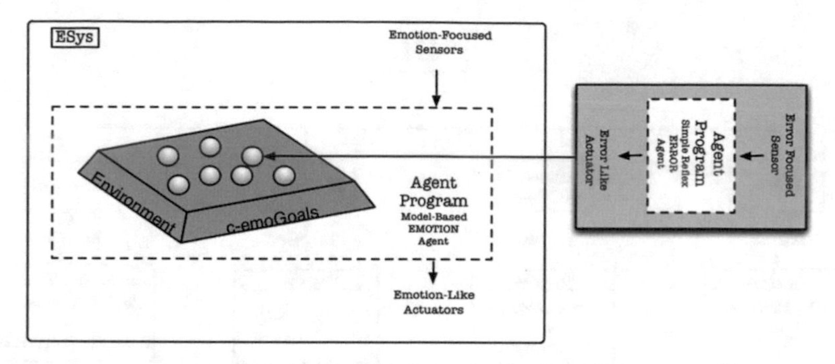

Fig. 5.27 An extended view of the agent based structure of an emotional goal and the relationship with the *Emotion Attribute* vectors

implement an equivalent topology in order to manage: (a) processes of error measurement, (b) processes of goals based dynamics, (c) processes related to memory requests, and (d) processes related to the emergence of new goals (see rationale ♣ 26). See Fig. 5.28(a) for a goal based model under SSADM (Structured Systems Analysis and Design Methodology) and Fig. 5.28(b) under SysML (the OMG Systems Modelling Language).

Thesis 13 ♣ *Four-role model of goals*
A fourfold-role model of *c-emoGoal*, *c-Concern* and *a-Dimension*, is conceived in an attempt to distinguish three different roles of behavior within each of them. The proposed formation is the subdivision of each artifact into four roles: (a) first-order objective: direct measurement of error, (b) second-order objective: relationship between error and memory resources, (c) a third-order objective: an associated dynamical system that evolves over time, and (d) four-order objective: the emergence of a new requirement of recovering error in the form of an emergent objective.

Each proposed subdivision needs to be detailed in order to obtain the whole meaning of this approach:

1. The direct measurement of error has been described within the work of an *Error-Reflex agent*.
2. The relationship between the emotional objectives and memory resources, defines a critical feature of the experience from the theorists' point of view. This will be detailed later in this chapter along the description of *'feeling-like'* operation.
3. The dynamical system associated with the specified item of goal accomplishment, which will enable the system for the fact of intending actions. This issue will be described in the following lines.
4. The emergence of new requirements is directly associated to the previous role. The modeling of dynamics enables new variables that might be used in order to reconfigure system requirements, system goals, or any other reconfigurable issue that needs to be considered. This role describes the required rules and models in order to compute reconfiguration at runtime.

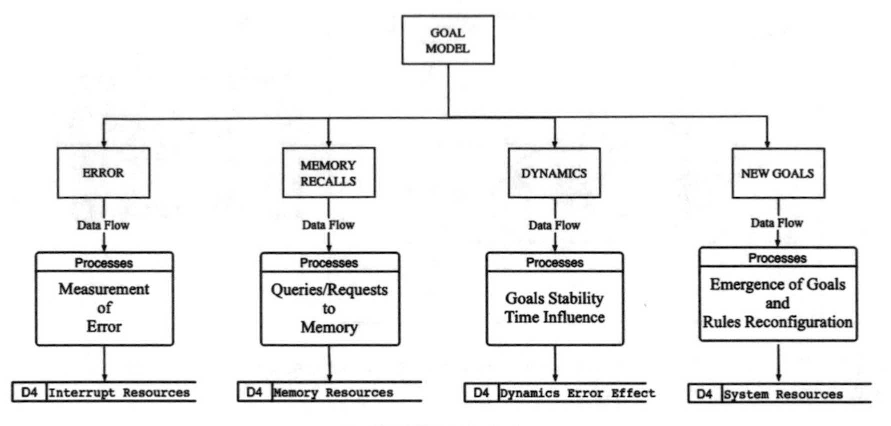

(a) SSADM based view

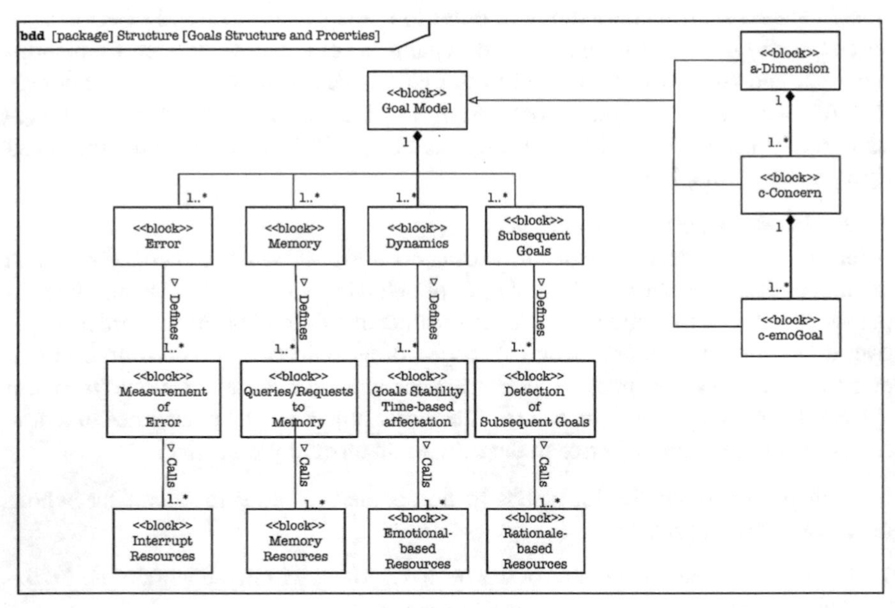

(b) SysML based view

Fig. 5.28 Structure of the system of emotional goals within an equivalent structure: **a** illustrates a view according to Structured Systems Analysis and Design Methodology (SSADM), and **b** illustrates a view according to SysML in order to clarify generalizations

IV. Goal constructs for dynamics As previously introduced, emotions are characterized by the widest range of dynamics for the systems. Sometimes, these dynamics are assumed to have the amount of physiological mechanisms associated to the situational state of the emotional goals.

However, maybe these are not the only dynamics provided to a system during an emotional response. Models should enable systems for the dynamic associated

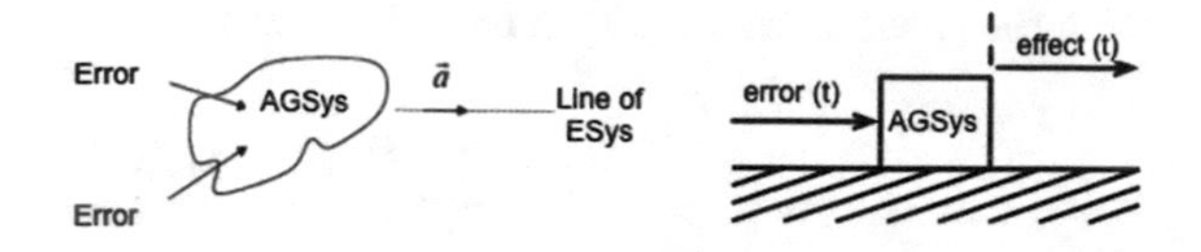

Fig. 5.29 Conceptualization of error cause-effect

to the action of intending something, related to the issue that causes the emotional behavior. This is an emotional-like dynamics associated to the emotion itself, as an effect of the design of the system. And this effect seems to derive from the behavior of the emotional goals.

What is expected regarding the emotional assessment is a qualitative analysis of error effects rather than the effects themselves. This does not mean that effects are not required, but that additional qualitative considerations are required. Our target is the use of this error in order to analyze relations between essential variables of change.

The aim is to transform the error in a difference that can influence the behavior of the system to deal with the error. That is, our objective is to transform this error in movement (without influence of inferences programmed by design) and let the system deal with this movement rather than the error. Freedom degrees in solutions are lower in inferences such as *'IF error THEN do'*, than in inferences such as *'IF error THEN power'*. In the second option, the system searches for different ways to *'do'* during the time in which this power is relevant in value (Fig. 5.29).

> Prior to continuing our description, we should remark that the following solution is an example of a possible solution in our approach. The differential equation herein selected aims at the most appropriate explanation of our approach. It is not argued to be the definite solution but a possible solution related to our approach.

We have conceived three artifacts (i.e. *c-emoGoal*, *c-Concern* and *a-Dimension*) in order to build those different schemes required to assess emotion from the view of an artificial system. In the following lines, we will analyze the form of these constructs in order to build artifacts that cause new effects related to the way this error influences the system, and that might be susceptible of being used in emotion assessment. Hence, the following model is generic to the three constructs—even if they will define the error within different levels of abstraction.

We will conceive the error—either regarding *a-Dimension*, *c-Concern* and *c-emoGoal*—as a displacement of AGSys[5] through the lines of the [wellness] reference that defines ESys. This allows us to conceive AGSys like a rigid system where Newton-like laws can be applied. That is, the error concerning each construct will be considered as a force that affects the system rather than a direct consequence for the system. Therefore, it allows an emotional analysis regarding how the error affects the system rather than the value of this error. Thus, we propose to make an analogy

[5]With our conceptualization, the displacement refers to some part of *AGSys* related to the error definition. However, for the purpose of an easier explanation, we will formulate our description by using *AGSys* and assuming what has been exposed herein.

with the Newton laws [258] in the following way:

$$\sum_i \vec{F_i} = M * \vec{a} \tag{5.13}$$

A relevant problem in artificial emotion design is that, somehow, emotion and behavior usually become a problem of direct cause-effect. Emotion is a dynamical like system, and the effects of an event do not occur immediately. From our perspective, emotion is a phenomenon that can be conceived as a dynamical system that evolves in time. We conceive this time evolution as dependent on the impulse that causes an error in either *a-Dimension*, *c-Concern* or *c-emoGoal*. With this conceptualization, we can assume that this error changes over time too.

This way, we can analyze the behavior of emotion by means of conceiving a dynamical system that models those main elements that concern the emotional evolution. These elements are (a) the error, (b) the own system and (c) the reference. For the sake of an easy explanation, we have chosen a classical mathematical model to explain our thesis. We have conceptualized that our error results in a force F that affects the *wellness of the system*. And that our system, concerning this error, will behave as a *Mass-Spring-Damper* trying to damp the effects of the error, and endowing the system with capabilities to restore its wellness. Nevertheless, this is just an example in order to describe our model. Each particular solution might describe any other model.

The error can be conceptualized as a force F that moves *AGSys* from the wellness line defined by *ESys*. From this perspective, the *cause-effect* relationship between the error—regarding *a-Dimension*, *c-Concern* or *c-emoGoal*—and the displacement caused by the effect of this error can be modeled by means of the three classical elements in *Dynamical Systems* [218]. That is, *mass, spring-restoring and friction* under the analogy showed in Fig. 5.30. This way, we enable the integration of the following key features in emotion science:

1. *Integrating variances from system to system*
   Even when the measurement of error might be the same for two different systems, the elements *mass, spring restoring and friction* are conditioning the effect of this error over the system. And it is additionally providing us a single method to model principles for *emotion tendencies* and *cognitive modulation of emotion*.
2. *Integrating energy because of its motion from wellness*
   Once the system is disturbed, it is not the error that produces emotion but the con-

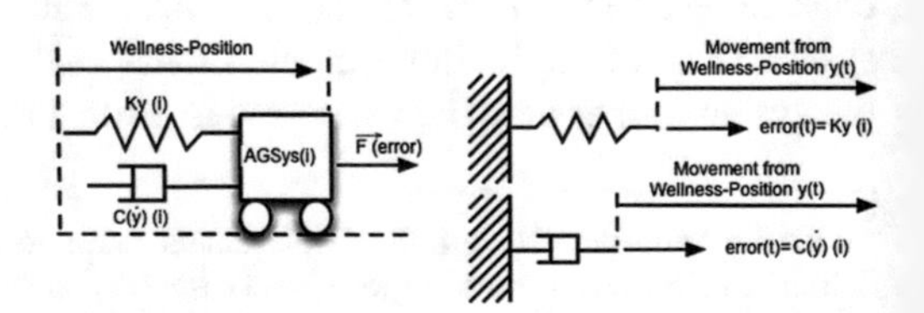

Fig. 5.30 Conceptualization of an emotional error damper system

sequences this error is creating within the system. The system becomes an unbalanced state of affairs. Biological systems result in a complex process by means of which symbolic representation and cognitive processes induce a requirement of wellness recovering, becoming the current objective that serves as a guide to behave. They are intricate processes in biology about what still has to be studied, and the identification of a starting point from which to create this new goal is not easy. However, artificial systems need this starting point, and we can identify it by means of the principles of energy in dynamical systems.

Now, our system AGSys and the effects that error produces in our system are analyzed from the viewpoint of the emotional goal. Here, attributes of the proposed dynamical system are defining how AGSys is affected by the error. And the effects over the dynamical system are used to assess the error effect. We thus have a *governing equation* that characterizes the dynamic state of the system, and that we can use in order to assess system state concerning emotion. We propose the use of this to calculate the response of the system to any error conceived as an external input $\vec{F}$.

The *mass* will be interpreted as a property of *AGSys* to store some type of energy that we will accept as *kinetic energy* concerning the error. And this kinetic energy is related to the movement that this error has caused in the system, which has moved AGSys from some wellness point. Generally speaking, kinetic energy is energy related to movement, and it can be associated to motion energy concerning wellness separation. As the system has moved from its wellness point, it has kinetic energy regarding the line of movement of wellness.[6] This way, we can analyze the emotion from the effect of the error over system and not from the error itself—which would translate into a direct cause-effect. Depending on the design, the value of *mass (m)* will provide different values of kinetic energy.

The *spring restoring* will be conceived as a property of *AGSys* to store some type of energy when being subject of the external force of the error, which we will accept as *potential energy*. From the perspective of physics, this energy represents the work a system has to conduct in order to move the system back to its original position (which is wellness, in our case). With this conceptualization, we additionally have the possibility of defining—or even reconfiguring from the basis of AGSys reasoning—the value of the spring constant *k*, allowing for emotional modulation and system tendencies (in order to model how the system faces each emotional situation). Finally, we will model the friction as a linear model [218] that will describe the linear relation between the error and the speed concerning wellness (since not all systems are affected in the same way concerning the same event and or the same error). Herein we are assuming that the two systems that are in touch are AGSys and ESys, since within our model, ESys is defining the line over which AGSys moves.

V. Reasoning the governing equation We will accept as a generality that *c-emoGoal*, *c-Concern* or *a-Dimension* are emotional goals. Hereafter, if we talk of *emotional goals*, we will thus refer to the complete set of these three constructs.

[6]Line of wellness is used for the sake of an easy explanation, since the proposed dynamical system allows for this simplification.

Our model represent the relationship between the error and the wellness displacement that AGSys suffers (input and output respectively). We will describe our explanation by assuming the continuity and the laws that characterize a dynamical system. We will thus assume that this model (i.e. this relationship) can be modified to work in discrete time domain. If variables are expressed as a rate per unit of time, the model can be expressed as a discrete time domain artifact.

We have conceptualized a model that will be a system under control, in a sense that AGSys will work to restore the wellness values. Hence, the system under control is our dynamical model that needs to maintain to a reference of null displacement (as we have assumed that initially the system is on wellness). We will assume however that AGSys will act on the original element from what error emerges, and no over the error itself. So that we do not need to describe the control loop in our model. That is, the *error(t)* will be conceived as a disturbance input generated within the *inner environment*.

With an *error(t)* as an external force acting on AGSys, we will obtain:

$$(AGSys)\ddot{y} + c\dot{y} + ky = error(t) \tag{5.14}$$

Regarding the displacement of our system from wellness, the variable of our interest is y. For some control system—that will be placed within AGSys, y will represent a control input. This y can be measured through computational artifacts. However, we will conceive y as a variable which we we will monitor in order to assess the stability of the emotional system. By assuming this model, we are modeling a system that behaves in a concrete way over time. And this behavior can be interpreted from AGSys in order to make evaluations about the effects of an error in some *c-emoGoal*, *c-Concern* or *a-Dimension* (we highlighten that they are conceived according to a three-role pattern, being one element an associated dynamical system).

The essential meaning of *'a valenced reaction'* is intrinsically associated with the concept herein defined *c-Valence*. We conceive *c-Valence* to capture the value of the positive or negative error concerning some *c-emoGoal*, *c-Concern* or *a-Dimension*. And from our viewpoint, it refers to the stability degree either in *c-emoGoals*, *c-Concerns* or *a-Dimensions*. And somehow, this is *qualitative information*.

We cannot monitor the entire dynamics of AGSys when an event occurs, and this is the reason why we introduce dynamics in our emotional goals. We can better design the emotional goal together with its own effect—and create a virtual system of effects, than monitor the complete AGSys searching for potential effects when an error occurs. Since the engineering definition of the emotional goal will be linked to an element of AGSys, the dynamic of the error effect will be always located.

We have conceptualized the dynamics of these constructs with a mathematical model of a mass-spring-damper system. This way, we can conceive an emotional error damper system concerning the influence of any emotional goal. However, we now must think about how to obtain qualitative information from the whole set of *c-emoGoals*, *c-Concerns* and *a-Dimensions* in order to characterize *c-Valence*. And this qualitative information should regard the behavior of each solution concerning the emotional goals.

By assuming an initial state of wellness concerning some emotional goal, we can specificy[7]:

$$\begin{aligned}
&(AGSys)\ddot{y} + c\dot{y} + ky = error(t) \\
&y(0) = y_0 \textit{ (Wellness initial position)} \\
&\dot{y} = v_0 = 0 \textit{ (Still in wellness position)}
\end{aligned} \tag{5.15}$$

And we can introduce the following state variables in order to obtain a first order model of this system:

$$\begin{aligned}
&(AGSys)\ddot{y} + c\dot{y} + ky = error(t) \\
&x(t) = \begin{pmatrix} x_1(t) \\ x_2(t) \end{pmatrix} = \begin{pmatrix} y(t) \\ \dot{y}(t) \end{pmatrix}
\end{aligned} \tag{5.16}$$

This way, our system can be expressed as a two dimensional and *first-order* differential equation:

$$\begin{aligned}
&\dot{x_1} = \dot{y} = x_2, \\
&\dot{x_2} = \ddot{y} = -\frac{k}{AGSys}y - \frac{c(\dot{y})}{AGSys} + \frac{k}{AGSys}error \\
&\dot{x_2} = \ddot{y} = -\frac{k}{AGSys}x_1 - \frac{1}{AGSys}c(x_2) + \frac{1}{AGSys}error \\
&\begin{pmatrix} \dot{x_1}(t) \\ \dot{x_2}(t) \end{pmatrix} = f(x_1, x_2, error)
\end{aligned} \tag{5.17}$$

Our emotional system is affected by the error, that is, the separation from the state of wellness. But it is also affected by the variation of this separation over time. They are two essential variables of emotional systems that we can describe in our solution. But now, what we require is not the value of these two variables, but some qualitative features that essentially might characterize emotion concerning the punctual emotional goal. This allows for feasible and successful solutions concerning the emotion state, with no requirement of explicitly computing the values of the state variables (i.e. the displacement from wellness and the variation of this displacement over time).

We have then characterized the behavior of our goal—concerning an error, as an *autonomous system* [218][8] with two state variables. And we thus have the option of representing these two variables in phase, where each point in the trajectory determines the state of the system at time t. Note that this system characterizes a

---

[7] We will assume in our explanation continuity in either function and partial derivatives, according to the constraints of C^1 [218].

[8] This concept of *autonomous system* is also defined by Boss [27] as a system whose second members are not depending on time.

concrete goal's behavior. And that it makes feasible an accurate characterization of an emotional state by enabling the possibility of defining equilibrium states and stability for wellness.

Regardless the model, the association of a goal with a dynamical system allows for:

1. The characterization of the effect of an error on state variables.
2. The assessment of equilibrium and stability concerning the state variables.
3. The reconfiguration of the model's attributes in order to obtain different *emotional tendencies* concerning the same error value.

In our example, our model represents the model that governs the *'movement from wellness state'* that AGSys experiences regarding an error in a punctual goal. This way:

1. We can characterize the effect of an error under two state variables that refer to: (a) the movement from wellness, and (b) the degree of this movement over time.
2. We can assure that this system is stable in the critical point of $(x_1 = 0, x_2 = 0)$, satisfying $(k > 0, c > 0)$.
3. We can change c and k in order to provide different *tendencies* since in our example, if $c > 0$, this results in a different asintotical equilibrium (this is a concrete result in our example; it will extend to different results depending on the considered model.)

As previously said, this reasoning about the *goverment equation* regards the model of any of the three essential constructs we are considering: *c-emoGoal*, *c-Concern* and *a-Dimension*. Later in this chapter we will conceptualize the way to make assessments. For now, our interest regards the growing of dynamics effects from bottom (i.e. *c-emoGoal*) to top (i.e. *a-Dimension*), and this is what will be analyzed in the following paragraphs.

VI. General concepts of emotional goals stability From our perspective, the emotion phenomena govern a whole set of inner processes in order to recover the system from unbalanced states. Unbalanced states refer to unstable behaviors regarding some feature of emotional goals. One of the essential features that our approach allows is the possibility of assessing the behavior of the error effects regarding stability. That is, we can analyze the effects of the same error depending on the initial conditions from which the state of a concrete goal starts.

As we argued in rationale ♣ 19, we will support the idea of a continuous process of emotion. Without regard herein to any complex consequence concerning this idea, it implies the possible influence of a new error before avoiding the effects of a previous error (i.e. the goal's *current state* is not the *wellness* based *initial state*). And small changes in those initial conditions might result in large changes in the state of the goal and maybe unstable states.

We will assume the characterization of the goal dynamics model as an *autonomous system* [218], by supposing that those trajectories [that feature the error effects] are depending exclusively on the initial [wellness] point and not on the initial time t_0 in which this error emerges. Even when it seems that t_0 might influence the emotional behavior of a system, we argue that it is not an effect of time but state.

The goal model of error effects can be characterized by means of models such as the one herein proposed (i.e. the *mass-spring-damper system*) by integrating the requirements that a concrete emotional theory might impose (regarding the effects of a punctual goal). As the dynamics of a goal effect will be initially described by means of an emotion-based theory, we can define these effects with the conceptualization of systems of differential equations [218] (trajectories regarding error effects, stability, critical points, …).

As previously described, we propose a taxonomy in order to describe emotional goals (see Fig. 5.22). We understand that an appraisal dimension (*a-Dimension*) determines a set of emotion-based concerns (*c-Concerns*), and that these concerns will define a set of goals accomplishment (*c-emoGoals*).

If *a-Dimension* regards the physical part of a generic artificial system *AGSys*, these final *c-emoGoals* will regard the physical part of *AGSys*. To give an example of this, we can refer to the *a-Dimension* of *"novelty"* defined by Ellsworth and Scherer [71]. This *"novelty"* regarding the physical part of *AGSys* could assess the *"happiness"* of *AGSys* regarding cue rises. Hence, *c-Concerns* will refer to classes of cues such as *sound based cues*, *touch based cues*, …and *c-emoGoals* will analyze the error regarding the amplitude of each cue within the class defined by a *c-Concern*.

Analogously, if *a-Dimension* regards the conceptual part of a generic artificial system *AGSys*, the final *c-emoGoals* will regard the conceptual part of *AGSys* allowing the exploration of rules. Under the principles of abstraction accepted in this thesis, this reasoning can be applied to assess within any level of abstraction. Once the hard problems of *symbolic representation* and *self* would be solved, this approach even should allow for assessing within the context of *social emotions*.

VII. General question about the feature of energy We have previously introduced some additional advantages regarding our proposal of modeling the effects of an error integrating dynamics. By assuming the same example of a *mass-spring-damper system* we will analyze these additional advantages. With this conceptualization, *AGSys* is attached to a spring-like effect (that can be modeled by the *Hooke's constant* "k") and a damping-like effect with constant c. We are interested in energy that might serve the system in the aim to recover a state of null error regarding wellness position. If we assume the loss of heat energy as part of *AGSys* configuration in order to restrict the amplitude of error effects, we can thus neglect this feature in our current reasoning.

With a null error, the position of *AGSys* regarding wellness does not change. However, if this error provides movement to our model, the position of *AGSys* is moved from wellness to some other y position. This way, the total energy $E(t)$ regarding this displacement will be the sum of the energy stored by the compressed or stretched spring and the kinetic energy of *AGSys*:

$$
\begin{aligned}
&(AGSys)\ddot{y} + c\dot{y} + ky = error(t) \\
&P_{energy}(t) = \int_0^y \mathrm{k} \cdot (\mathrm{err})\, \mathrm{d}(err) = \frac{k}{2} \cdot y^2 \textit{ (Potential Energy)} \\
&K_{energy}(t) = \frac{AGSys}{2} \cdot (\dot{y})^2 \textit{ (Kinetic Energy)}
\end{aligned}
\tag{5.18}
$$

With an error:

$$
E(t) = P_{energy}(t) + K_{energy}(t)
$$

Thus, by using the values of $E(t)$, $P_{energy}(t)$, and $K_{energy}(t)$, we can represent additional values that allow for some type of "emergent system requirements of wellness recovering". This way, we can use the following vector of features in order to allow for emotional assessment:

$$
\overrightarrow{V_{assessment}} = \{x_1; x_2; E(t); K_{energy}(t); P_{energy}(t)\}. \tag{5.19}
$$

VIII. Generalization As was introduced in rationale ♣ 27, the use of differential equations is aimed to provide information previously unknown about the interaction of goals, and to analyze their qualitative effects. The use of this mathematical artifact is proposed in order to provide knowledge about goals behavior and relationships among goals. Therefore, the stability comes from the equilibrium in the interaction of these goals. And it is an essential informational artifact in order to provide emotional meaning (see Fig. 5.31)

The generalization of this idea regarding *c-Concerns* and *a-Dimensions* is made on the basis of an analogous reasoning. This time, however, it is necessary to remark the different nature of *c-Concerns* and *a-Dimensions* compared to *c-emoGoals*. Any bounded set of *c-emoGoals* are defined in order to accomplish a concrete meta-goal of a *c-Concern* (and this conceptual operation is repeated for *a-Dimensions* and *c-Concerns*). Thus, the differential equation that governs a concrete *c-Concern*

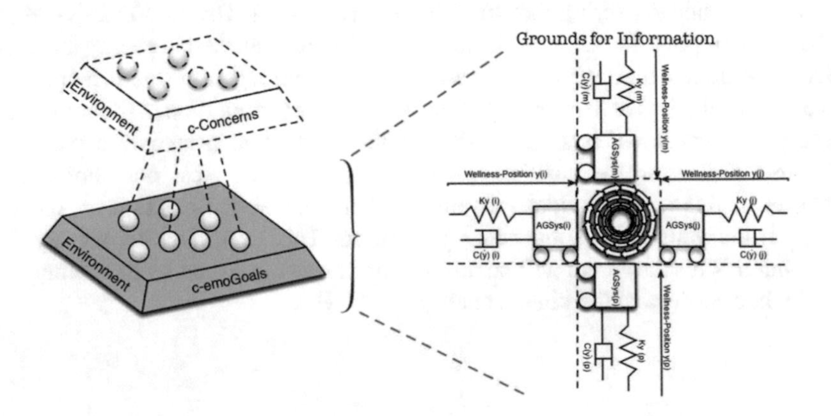

Fig. 5.31 Quality of c-Value

is related to the accomplishment of certain objectives related to the *c-emoGoals* (analogously for *a-Dimensions*). And these objectives may be related to stability, relationships, behaviors, ..., concerning *c-emoGoals*. Hence, the differential equations that govern a *c-Concern* and *a-Dimension* are not the same for all three types of artifacts (see rationale ♣ 25).

This way, generalization refers to the procedure that regards the hierarchies of the pyramidal structure, not to the agents that work as units within the structure.

5.6.3 Valenced Reactions

We have characterized the behavior of the individual agents that define the structure of the emotional goals. And that it makes feasible an accurate characterization of an emotional state by enabling the possibility of defining equilibrium states and stability for wellness.

As we analyzed in Chap. 4, we refer to *c-Valence* in an analogous sense to that expressed when *'valenced reactions'* are argued in emotion science (see rationale ♣ 23). It is a different concept from that of *v-State* that refers to a more high level state that integrates relationships. This *c-Valence* is used to capture the value of the positive or negative error concerning some *c-emoGoal*, *c-Concern* or *dimension*.

AGSys can deploy valenced reactions to the consequences of an event, agent or object (by referring to the *OCC model* as an example [199]). And the valenced reaction regards either *c-emoGoals*, *c-Concerns* or *a-Dimensions* by capturing their error.

In our example, our model represents the model that governs the *'movement from wellness state'* that *AGSys* experiences regarding an error in a punctual goal. By making use of the vector of features that has resulted from our analysis:

$$\overrightarrow{V_{assessment}} = \{x_1; x_2; E(t); K_{energy}(t); P_{energy}(t)\} \tag{5.20}$$

The effect of an error in x_1 and x_2 refers to: (a) the movement from wellness (x_1), and (b) the degree of this movement over time (x_2). They are two initial values that can be used to allow *AGSys* to be aware about some relevant change. And *AGSys* should trigger some type of process in order to obtain improved information about this change (see thesis proposal 61). If these processes regard artificial systems, they commonly respond to *interrupt processes* in order to reorganize system resources aiming to face critical states.

Herein, some essential processes that should be triggered in case of emotion design are described: (a) *attentional-like* processes in order to improve the environmental information in critical states, or (b) *action readiness* processes aimed to reconfigure the normal work of the system to another of higher alertness and preparation for action. The ability to improve survivability will depend on this essential capability of readiness.

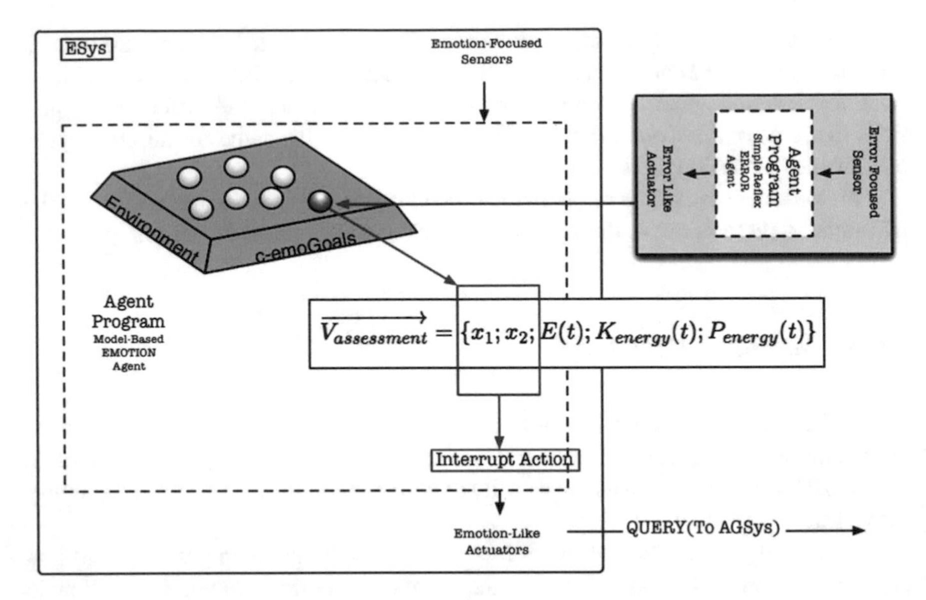

Fig. 5.32 Conceptualization of a state of action readiness query, which comes from a concrete c-emoGoal

This thesis proposes the use of x_1 and x_2 as the source for triggering those actions of interruption. Since they are the initial detectors of movement from wellness state (x_1), and how this movement is evolving over time (x_2), they become acceptable options in quality in order to trigger [interrupt actions of] readiness (see Fig. 5.32).

Thesis 14 ♣ *The question of Action Readiness*
When we refer to artificial systems, *alertness* and *action readiness* result in the need for reconfiguring the normal work of the system to another of higher attention and preparation for action. This depends on how to be aware of the current context regarding current needs of the system, deal with this context, and reconfigure the space of the problem for success. This thesis proposes the use of x_1 and x_2 as the source for triggering those actions of interruption, so that *c-Valence* is assumed to have the following form:

$$c\text{-}Valence = \mathcal{F}(x_1, x_2). \tag{5.21}$$

5.7 Feeling in AGSys

We have already conceptualized the different key terms in emotion science aiming to allow for computational comprehensibility. Nevertheless, the real interest of artificial emotion emerges from the core of the symbolic representation of *feeling*. Regardless agreements or disagreements about the argument of Damasio:

> We know that we have an emotion when the sense of a feeling self is created in our minds. (...) we only know that we feel an emotion when we sense that emotion is sensed as happening in our organism.
>
> Damasio [54]

The essential problem in engineering is the nature of this representation and how artificial systems might become aware of their own emotions. The problem of feeling is intrinsically joined to the problem of symbolic representation. This is one of the most characteristic features of human beings by means of which, sensorial information is transformed into a higher abstract form that afterwards is used to reason.

This thesis addresses the problem of how the *feeling* might become a *sense*, and how it might be symbolically represented. That is, how to transmit the feeling to the system reasoning so that *AGSys* can acquire signification by association with its situational state.

I. The approach to the problem As previously introduced, the analysis made in this thesis regarding representation is supported by the theory of Damasio. This work is probably the one that can be better used to computationally explain the process. Hence, this thesis will describe those principles of representation postulated by this author. Principles to build a representation of computational feeling by using the requirements concerning *self*.[9]

Earlier in Chap. 4 the four requirements in order to represent emotion were analyzed (see rationale ♣ 44): (a) How to induce emotion, (b) what the essential stimuli in order to induce emotion are, (c) how emotion is measured and categorized, and (d) how the systems can feel emotions. The theory of Damasio frames this problem within the essential supporting structure of four coarse grain processes:

1. Basic Regulation
2. Emotion
3. Feeling
4. High Reasoning (by means of which *feeling* is experienced)

II. General assumtions The meaning of an *inner-Object* from the perspective of this thesis has already been described, as well as how this artifact is represented for the purposes of this work. This abstraction has been assumed as the analogy to the object of Damasio, which establishes the starting point for his theory. The perception of an external object with this abstraction will thus be assumed, together with the feeling regarding this object under the effects that this abstraction causes within the organism.

The representation of Damasio within the framework of this thesis is described based on the sequential growth of this *inner-Object*. This growth is developed by

[9]The search for possible solutions that might help to solve the hard problem of *Self* is out of the scope of this thesis. Instead, this work just uses those principles described in a widely accepted theory in order to build a computational *feeling*.

means of the integration of information that regards the emotional assessment concerning the effects caused by the object. This way, this *inner-Object* becomes a grown object named *emotion-Object* (by doing an interpretation of the arguments of Damasio—see rationales ♣ 40 and 48 in Chap. 4)

The statement of *'subjective awareness'* strongly communicates that each system uses its own references (uniquely established within the system) and makes its own representations. Any external object that innerly becomes an *inner-Object* will interact with perceptual stages where lots of interoceptive processes are triggered. Two systems perceiving the same external object will thus infer differences in the qualitative nature of the object. The *'subjective awareness'* will endow different meanings to the *inner-Object* depending on the system where it is conceptualized (see rationale ♣ 45).

Therefore, as we have advanced lines above, the broad sense in which Damasio uses the term *object* allows us to conceptualize *emotion* and *feeling* as *inner-Objects*. And additionally, as Damasio distinguishes between *feeling* and *emotion*, we will thus distinguish between *emotion-Object* and *feeling-Object*. And the conceptualization of a *feeling-Object* will be conceived in an analogous form as Damasio conceives the *feeling* before *'knowing feeling'*.

Consequently, this thesis aims to build the *feeling-Object* as an evolution of the *emotion-Object*. As this *inner-Object* can grow in abstraction by definition, it will represent firstly emotion (*emotion-Object*) to later represent feeling (*feeling-Object*). Thus, *emotion-Object* and *feeling-Object* will differ in 'gain' regarding 'abstraction-degree'. Furthermore, this conception enables subjectivity integration since, by managing the same *emotion-Object*, each system might obtain a different *feeling-Object* (see Fig. 5.33).

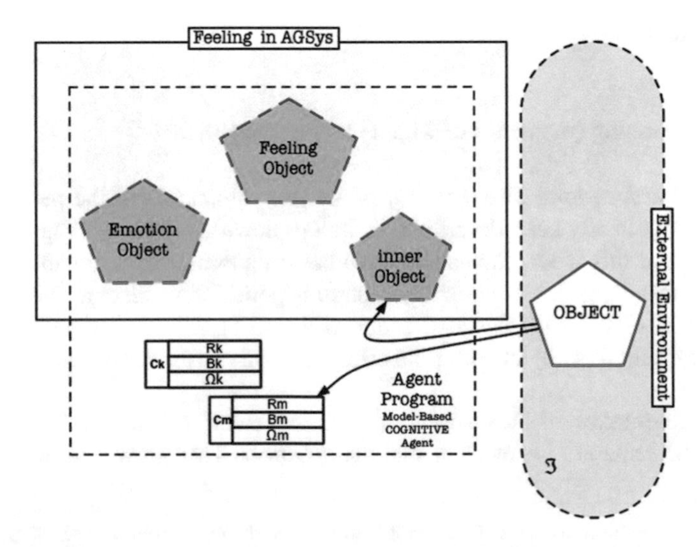

Fig. 5.33 Conceptualization made in this thesis regarding the threefold representation of an external object: *inner-Object*, *emotion-Object* and *feeling-Object*

5.7.1 *The Problem of Representation in AGSys*

This analysis describes essential principles in order to build the representation of computational feeling, by using the requirements postulated by Damasio concerning *self*. Once conceptualized the relationships required to build this *feeling-Object*, additional features regarding memory will be assumed as future works.

The same limitation is found regarding the problem of *Self*. The foundations that exist behind this concept are out of the scope of this thesis. This thesis just assume the characterization of *AGSys*, in such a way for it to be capable of exploiting models related to memories during the emotional response.

As introduced in the previous chapter, the concept of *biographical memory* constitutes an essential element concerning the emotional experience. It is argued to be involved in the experience of *self*, characterizing the phenomenon of representation (see the work of Conway and Pleydell-Pearce).

The idea of memory related to life systems, constitutes the amount of retrievable data aiming to keep essential information for future use. Related to artificial systems, this memory however refers to both the place and the data stored for short or long term retention. From the perspective of Conway and Pleydell-Pearce, the distinctive feature of the *biographical memory* is the attribute of *self-referring*. And they hold that *biographical memory* is "transitory dynamic mental constructions generated from an underlying knowledge base" [50].

I. General overview of the theoretical analysis

This is a summary of Sect. 4.6.1 in Chap. 4.

Self-Memory System (SMS): refers to the conjunction of the *Working Self* with the *Autobiographical Knowledge Base*, and it is conceived of as a superordinate and emergent system (see Sect. 4.6.1 in Chap. 4 for additional explanation).

Conway and Pleydell-Pearce explain the *autobiographical-Knowledge* within a structure of three broad levels and featured by the attribute *Self Referring*. These three levels are named: (a) *lifetime periods*, (b) *general events*, and (c) *event-specific knowledge (ESK)*.

1. Lifetime periods.
2. General Events: Feature vivid memories of events related to the accomplishment of personal goals.
3. Event Specific Knowledge (ESK): Predictor of memory specificity, defining feature of memory vividness and responsible of linking imagery with General Events. And finally, an essential feature of ESK is that they are central to *autobiographical remembering*. Early versions of the theory refer to ESK and later versions to *Episodic Memory*.

The perspective of Damasio about *feeling* and *emotion* is the following: any change in our organism, is mentally represented by means of neural patterns and resulting images. And if those images are occurring with a sense of *self* one instant later and are enhanced, they become conscious. This 'one instant later' justifies the validation of those links to recover *General Events* from the imagery in *ESK*.

Regarding the relationships between the memories and the *self*, Conway and Pleydell-Pearce provide the following conceptualizations related to the *Working Self*:

1. The Working Self: Used to make a connection with the *Working Memory* (which integrates processes aimed to controlling separate systems).

2. The goals of the *Working Self* are assumed to be a subset of the control processes integrated in the *Working Memory*, and they constrain *cognition* and *behavior*.
3. The goals of the *Working Self* are essential artifacts in order to control self-discrepancies in the *Autobiographical Memory*.
4. Autobiographical Memory: Limits the range and types of objectives that (within healthy states) can be hold.

Regarding the *Working Self Domains*, three separated domains are conceptualized:

1. Actual Self: Accurate representation of one's self.
2. Ideal Self: Represent the aspirations.
3. Ought Self: Rules learnt through life.

And "discrepancies among the three domains lead to characteristic forms of negative emotional experience": They are argued to be solved by means of *feedback control loops* where inputs represent the state of the world of discrepancies, and the control aims to reduce these discrepancies.

Regarding the *Working Self and General Events*: It is argued that General Events arise in response to experiences in which self (i.e. working self) and goals (i.e. working self goals) are highly integrated or strikingly disjunct.

II. Relationships between Emotion, Memories and Self

This is a summary of Sect. 4.6.1 in Chap. 4.

Conway and Pleydell-Pearce argue that "vivid memories often arise in response to experiences in which the self and goals were highly integrated (...) or strikingly disjunct".

And they hold that "the extent to which individuals were able to effectively use appropriate cognitive reactions to deal with dissonant memories was positively related to their sense of well-being, suggesting that control of memory may have far-reaching implications for mental health".

Putting the idea of Conway and Pleydell-Pearce in computational terms, *emotional memories* can be viewed as computational artifacts whose attributes are *knowledge of goals* and *emotion experience*. First of all, the concepts of *emotional goals* and *working-self goals* should be separated. It seems as it they are referring to different realities. Secondly, and even disregarding nature, from the knowledge of goals to the emotional experience there should be some methods that relate these two types of attributes. Conway and Pleydell-Pearce propose two models as the base to build these methods: (a) the model of Carver and Scheier [44] that relates goal-attainment abandonment and emotion, and (b) the model of Schemes [233] that relates the functions of maintenance, repair and change goals and emotional experience.

Conway and Pleydell-Pearce describe the *action-guidance system* [44] as a model of the relationship between the abandonment of a goal-attainment and emotion. And that this system is monitored by a second system to *assess and modulate the rate at which the goal system reduces discrepancy*.

This second system is argued to be *the emotion system, and positive emotions reflect an acceptable rate of discrepancy reduction, whereas negative emotions reflect an increasing failure to reduce discrepancies*.

The last consideration is essential regarding this thesis: this second system is argued to be the work of the emotional system. Analogously, positive emotions reflect an acceptable rate of discrepancy reduction, whereas negative emotions reflect an increasing failure to reduce discrepancies.

> The *Communicative Theory of affect* by Conway and Pleydell-Pearce becomes a perspective where goals and plans "communicate with other processes and structures by means of their output, and other parts of the cognitive system communicate with the goal only via its input, as in the negative feedback loop".

Even if we presume high complexity concerning this negative feedback loop, the final idea it motivates is the control of goals and associated plans under some complex set-points focusing on the maintenance of their stability. When changes in the probability of achieving an essential goal are detected, the monitoring mechanism transmits alert signals to the broad cognitive system that "sets it in readiness to respond".

> The argument found in this work that assumes "By this communicative theory of affect it is the alert signals from the monitoring mechanisms that are experienced as emotions" [50], is surprisingly in accordance with the controverted theory of James [118] (subscribed, accepted, discussed and denied), who establishes the origin of emotions at the sense of bodily signals.

III. Self memory system Our objective is to conceptualize those artifacts that any *AGSys-Self* would require, based on computational models and the interaction of those models. Principles of design will be derived from those principles described in the theory of Conway and Pleydell-Pearce [50] analyzed in the previous chapter.

Some conceptualizations postulated by authors (i.e. semantic, episodic, procedural memory and so on) are not so much the current focus, but rather left for the future works derived from this thesis. The objective is to propose some principles and foundations under which a computational *Self-Memory-System (SMS)* might be delineated, in accordance with the approach of this thesis (Fig. 5.34).

The *Self-Memory-System (SMS)* will thus be assumed to be a system that arranges pre-stored knowledge into a form which can be exploited as a memory by *AGSys* (see rationale ♣ 32).

The authors describe a model of autobiographical memory in which memories are transitory mental constructions within a *Self-Memory-System (SMS)*. The *Self-Memory-System (SMS)* contains the *Autobiographical Knowledge Base* and current goals of the *Working Self*.

Within the *Self-Memory-System (SMS)*, control processes modulate access to the *Autobiographical Knowledge Base* by means of a successive shaping cues used to activate its structures. This way, specific memories are formed. The relation of the *Autobiographical Knowledge Base* to active goals is reciprocal, and the *Autobiographical Knowledge Base* "grounds" the goals of the *Working Self*.

IV. Autobiographical Knowledge Base The conceptualization of the *autobiographical memory* allows us to provide a computational means in order to establish essential foundations for mental-like processes (see rationale ♣ 32).

The architecture of both the data and the storage system are essential blueprints that influence the final use as an informational system. Concerning the purposes of this thesis, this informational system should provide the means to obtain an essential part of the emotional experience.

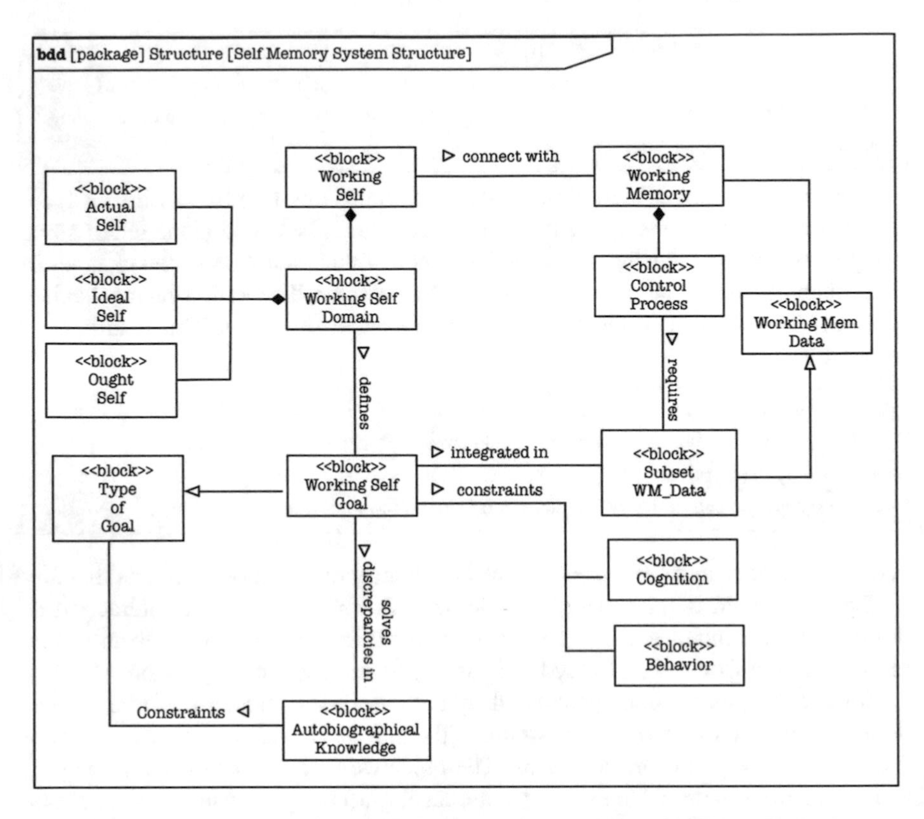

Fig. 5.34 Generic taxonomy of the Self-Memory-System under the OMG SysML that integrate main concepts and properties of Conway and Pleydell-Pearce theory [50]

As *AGSys* is conceived as a knowledge-based system, it is consequently enabled to represent *abstract objects* under the perspective of Gruber [98]. This characterization has been already used in order to build the *inner-Object*. Applied to the current problem of memories, it results in the capability of *AGSys* to build series of representations—in the form of *chunks of memories*, that can be afterwards related to the perceived *inner-Object*.

Definition 5.19 ♣ *chunked-Memories*
Computational artifact that we situate inside the system which represents solid pieces of memories as a source of knowledge for the system.

In order to address our computational objective, we thus need to make all these conceptualizations more clear. Hence, a process by means of which to build these chunks of memory is required, so that they an be related to the emotional event.

We assume the characterization of *AGSys* in such a way that it should be capable of exploiting autobiographical knowledge related to emotional events. We thus strive for the use of the same computational foundation as the one argued for the construction of the *inner-Object*: the use of ontologies to build knowledge based systems (see works by Bermejo-Alonso [21–23]).

Figure 5.35a describes a generic taxonomy of the *Autobiographical Memory*, (b) describes the concept of *General Event* and (c) shows the conceptualization made for the purposes of this thesis of a possible *ESK* (a.k.a. *Episodic Memory*). Refer to rationale ♣ 35 in order to obtain a better comprehension of these illustrations.

V. Episodic memory Nowadays, the *Episodic Memory* is considered a separated system whose representations feature an amount of characteristics, among which there is the retainment of summary records related to sensory–perceptual–conceptual–affective processes derived from *Working Memory* (see Sect. 4.6.5 in Chap. 4). They are roughly represented, and this representation defines short time slices of visual images determined by changes in goals processing, which are durable just if they are linked to the *Autobiographical Knowledge*. And they are recollectively experienced when accessed.

As argued by rationale ♣ 36, the understanding of the role that *Episodic Memories* play concerning emotion is essential. However, the design of analogous computational models is part of a hard problem that defines still open fields of research beyond engineering.

For now however, it could be modelled by means of those artifacts that current artificial systems enable. Regarding the summary records, a single piece of information coming from the context, together with specific patterns of activation can be wrapped into the artifact of an object with partial information. That is, an object which integrates partial information from different sources, which is regulated by a concrete control process in order to be stored only when required, and which can be recalled by any event that might affect these sources.

5.7.2 Regarding the Problem of Autonomy in AGSys

Autonomy is commonly referred to as the quality of systems to independently behave while pursuing mission objectives [106], and it is widely recognized to still present important open issues. With the common perception, *autonomy* seems to refer to the capability of thinking and consequently acting for itself, rather than just being free from external control.

From my own perspective, one essential missing point in control is the lack of the *Working-Self Goals* as critical features of independence (see rationale ♣ 34). It seems that *Working-Self Goals* might be closely associated with those artifacts defined in this work as *a-Dimensions*, *c-Concerns*, *c-EmoGoals* and *Emotion Attributes*.

We argue that one key feature of *autonomy*—which consequently will affect the control for *autonomy*, is the structure and management of the *Working-Self Goals* structure. We hold that control of discrepancies within this structure, is a key aspect in *control for autonomy*. The operation of managing *Working-Self Goals* comes to

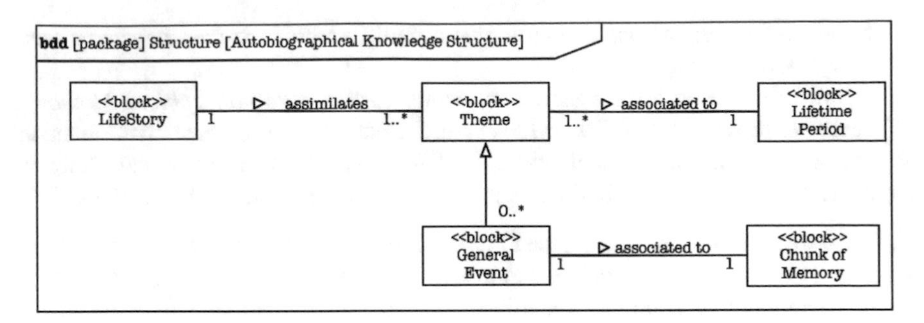

(**a**) Autobiographical Memory KB

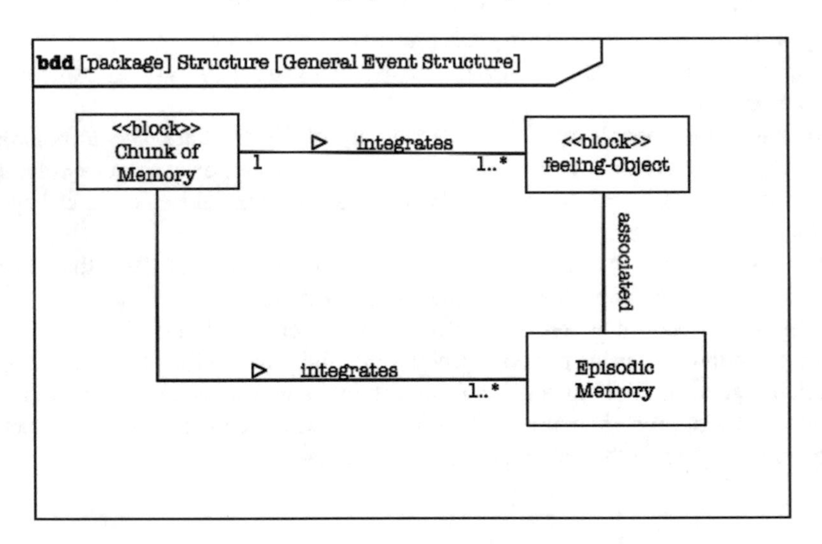

(**b**) General Event

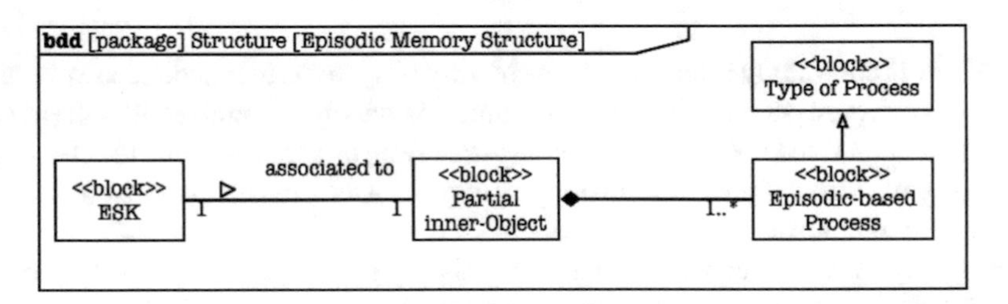

(**c**) Episodic Memory

Fig. 5.35 Taxonomy of a generic Autobiographical Memory Knowledge Base: **a** Classification of concepts according to the theory of Conway [49] (Fig. 4.4 shows the conceptual view of the authors), **b** Conceptualization of a General Event structure, and **c** Conceptualization of an Episodic Memory structure

solving discrepancies when external events appear, and these external events might concern either environmental changes or commands to complete tasks.

Thesis 15 ♣ *The problem of Autonomy in AGSys*
One key feature of *autonomy*—which consequently will affect the control for *autonomy*, is the structure and management of the *Working-Self Goals* structure. Control of discrepancies within this structure, is a key corner in *control for autonomy*. The operation of managing *Working-Self Goals* comes to solving discrepancies when external events appear, and these external events can regard environmental changes or commands to complete tasks.

Since *AGSys* is assumed to be a Model-based cognitive agent, the interpretation of this scenario can follow the approach of Vogel and Giese described within Sect. 2.7 in Chap. 2. That is, the work of the *Self-Memory-System* can be considered under the principles of an adaptive engine (i.e. feedback loop) that integrates the operation of the *Working-Memory* under the constraints of the *Working-Self Goals* (see Fig. 5.36).

I. Reflection model The *Working-Self Goals* will constitute the reflection model (i.e. the common shared knowledge) of an adaptive engine that monitors the state of the *Emotion Attributes* in order to constrain behavior and cognition by means of adaptive actions. This reflection model is updated by means of a higher adaptive engine in order to allow for those limits that are argued to be provided by the *Autobiographical Memory* on the type of *Working-Self Goals*.

For now it will be assumed that for the same *AGSys*, for any update made on the reflection model, this update will not affect the number of *a-Dimensions* assumed for *AGSys*. This assumption is made in order to prevent misunderstandings regarding updates.

II. Monitoring The monitoring models observe and check the progress of the components of the *Emotion Attributes*. The observation of these components and the abstraction of *Working-Self Goals* are mapped in order to prepare the following analysis.

III. Analysis The state of these *Emotion Attributes* is analyzed based on evaluation models that define (a) the requirements of well-being for the states of *a-Dimensions*, *c-Concerns* and *c-emoGoals*, (b) the requirements in order to avoid discrepancies regarding the *Autobiographical Memory*, and (c) the related constraints on the reflection model of *Working-Self Goals*.

IV. Plan The revealed state is analyzed based on change models in order to decide on the entire set of discrepancies (either regarding well-being or related to the autobiographical knowledge). This phase will describe the space of variability that implements *AGSys* in order to self-adapt concerning its situational emotional state.

V. Execute The execution models will refine the obtained model-level adaptation (which comes from the previous stage of planning), and carries out the course of actions regarding the *AGSys* level adaptation. This way, the different levels of abstraction from which this solution can be assumed must be clarified. This action

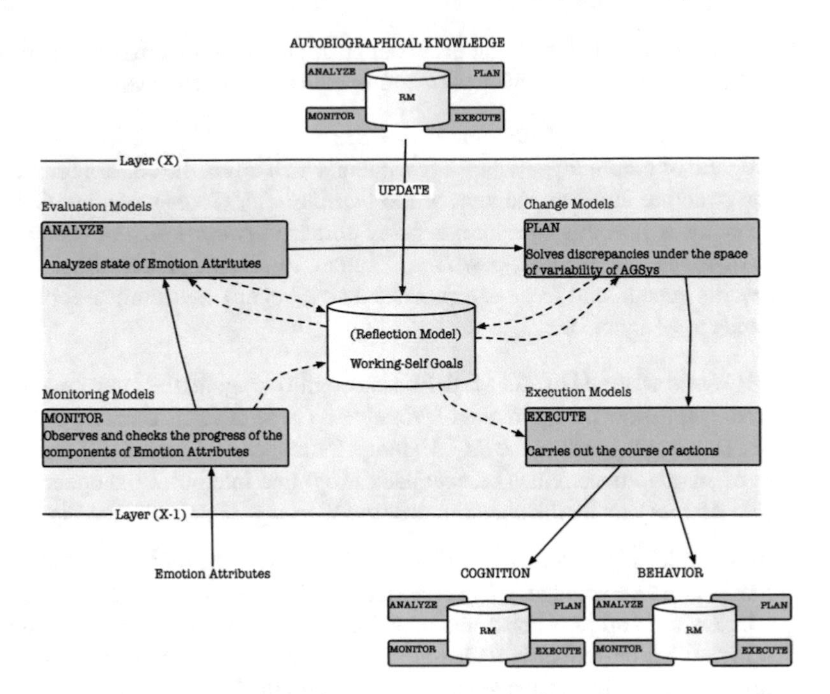

Fig. 5.36 Conceptualization of the *Self-Memory-System* operation by using the work of Vogel and Giese

of execution is understood herein as meta-actions in order to influence *AGSys* levels of adaptation regarding behavior and cognition. And the problem of imagery and related chunks of memory will become an issue to be dealt with from the perspective of the *AGSys* levels of adaptation regarding cognition (i.e. since it is related to the problem of perception).

5.7.3 The R-Loop

The *inner-Object* has been conceptualized as the computational artifact situated inside *AGSys*. This artifacts aims to represent descriptive aspects of an external object, in order to enable *AGSys* for its exploitation as knowledge. This *inner-Object* is considered as the first form that an external object acquires inside *AGSys* in any given perceptual process. And it is argued to be built by means of ontology-based tools.

Regarding emotion, this *inner-Object* is just an artifact aimed to represent the external object. And once perceived in this initial form, it causes influences on *AGSys* that are critical for emotion perception. This thesis assumes that the emotional assessment is an operation by means of which the quality of *value* is perceived. And 'quality' is herein assumed as an increased abstraction, built on the source of a measurable difference.

After studying how *AGSys* might build and integrate this *value*, the need for this *inner-Object* to grow regarding both quantity of information and abstraction degree has been presented. And the solution proposed is the conceptualization of this *inner-Object* as an argument of a *perceptual recursive function* (see Fig. 5.16).

This thesis assumes the conception of a topological type of the perceptual function, and the design of the inner-Object with such a topological type so that it will not change this topology over time. And the topology of an *attractor set* has been proposed for consideration and discussion.

I. Splitting the inner-Object There is common agreement that emotion and feeling are two separated concepts. The way such assertion affects this work is that emotion and feeling cannot be conceived within the same domain of the problem. Regarding emotion, the *inner-Object* will cause interactions from which emotion and feeling will emerge within *AGSys*.

Along the explanation of feeling, [54] questioned the perception of emotion—regarding feeling, as a new object that comes from inside the system. This question brings about new relevant issues for this thesis.

It can be assumed that information coming from the environment can be conceptualized in the form of an *inner-Object*. This object results in another object that represents the emotion once the emotional effects of the former *inner-Object* are processed. And with respect to the building of feeling, this new object built on the basis of the emotional effects will result in an analogy of the external object.

And this is in accordance with the argument of an *inner-Object* that recursively is growing and integrating information: the *inner-Object* grows to become an *emotion-Object* (that integrates the former), and *emotion-Object* grows to become a *feeling-Object*.

Perception is a complex process full of processes that provides a complex view full of information and relationships. And emotion is just a part of this entire process of perception. From this perspective, the *inner-Object* should be splitted into two separated projections inside *AGSys*.

The *inner-Object* will thus represent the conceptualization of the object regarding its shape. And the first object for the emotional assessment will be defined by the conceptualization of the first emotion-based effects that it causes inside *AGSys*.

This new object has been previously defined as the *emo-inner-Object*: a computational artifact that we situate inside of the system, and that represents those relevant changes that an external object causes on *AGSys* concerning emotion-based references (see Fig. 5.37).

Recalling essential definitions proposed previously in this chapter:

Definition 5.20 ♣ *emotion-Object*
Computational artifact that we situate inside of the system, and that accumulates the information about the consequences that an external-Object is inducing inside of the system.

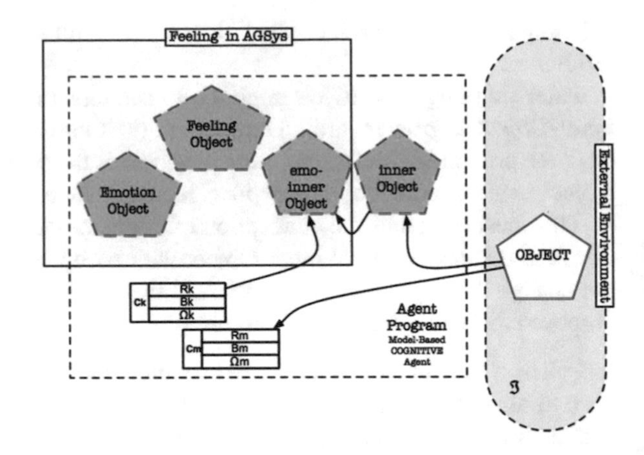

Fig. 5.37 Conceptualization of the three assumed artifacts in order to approach the experience of emotion

Definition 5.21 ♣ *feeling-Object*
Computational artifact that we situate inside of the system, and that represents an evolution of the *emotion-Object*.

Consequently:

Definition 5.22 ♣ *Computational Process of Emotion (cp-Emotion)*
Process by means of which the emotion-Object is built.

Definition 5.23 ♣ *Computational Process of Feeling (cp-Feeling)*
Process by means of which the feeling-Object is built.

The *v-State* has been conceived as a midway step that acts as an intermediate stage between *emotion-Object* and the *feeling-Object*. This solution aims to provide a computational bridge between two artifacts that come from two different domains (which consequently influences the way they are described). Figure 5.17 illustrates this idea.

However, this formulation can be generalized for a general application in order to provide the conceptual bridge between the *emo-inner-Object* and the *emotion-Object*.

II. R-Loop conceptualization The *recursive-Loop (R-Loop)* is a consequence of the question about how the *inner-Object* evolves, and what is the form of this recursive function that integrates new knowledge within the object at each recursive step.

R-Loop is conceived as the recursive functional artifact, aimed to provide an explicit interpretation of the object by means of successive executions of the same conceptual loop. Each recursion is conceived within the work of a conceptual loop, which performs the analogous function at each repetition but having a different object as the origin.

Definition 5.24 ♣ *R-Loop*
Recursive functional artifact, aimed to provide an explicit interpretation of the object by means of successive executions of the same conceptual loop.

The unit loop's performance at each recursion is based on the theory of Damasio. It is related to the integration of those relationships that emerge from the interaction between the object and the system. And this thesis describes a computational interpretation of this approach presented by Damasio, which will be detailed in the following section with the name of *Model of Emotion Perception (MEP)*.

By introducing the performance of this loop, it can be described as an integration of the following artifacts: (a) what may be called the 'knowledge of the object', (b) the effects that this object causes on the system, and (c) the relationship between the two previous artifacts. The result is a new object that integrates the previous object, together with new information that regards interactions with the system (see Fig. 5.38)

It should be noted that each concrete *R-Loop* might be designed aiming at different perceptual purposes. Consequently, the study about these artifacts previously mentioned will be related to each perceptual purpose. This thesis will address purposes of emotional assessment.

5.7.4 MEP Model

Prior to continuing our description, it is essential to emphasize the importance of what herein is understood by *c-Map* and *c-Image*. They represent the computational analogies of those *maps* and *images* to which Damasio refers.

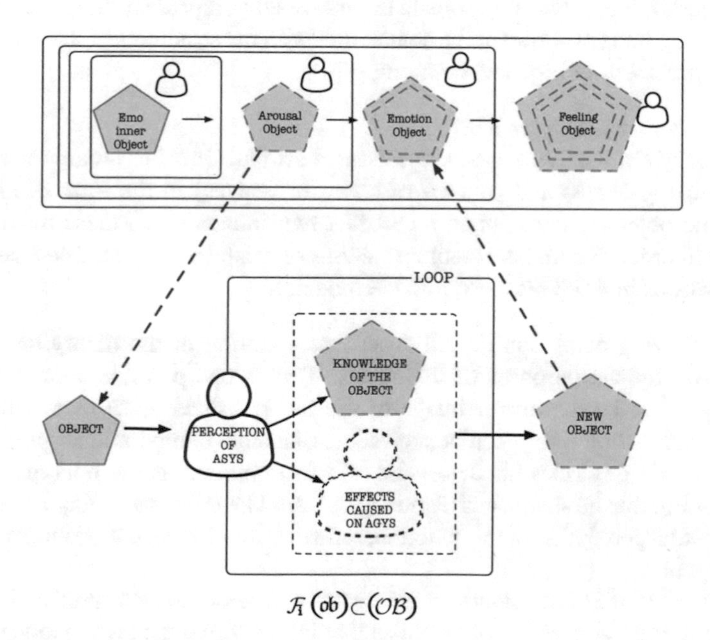

$\mathcal{F}_1(ob) \subset (\mathcal{OB})$

Fig. 5.38 Conceptualization of an operative unit in *R-Loop*

♣ *computational-Map (c-Map):*
It will refer to patterns of relationships concerning a set of system artifacts. They are patterns that (a) are not conscious for *AGSys*, and (b) serve to build *c-Images*

♣ *computational-Image (c-Image):*
It will refer to patterns of states concerning a set of system artifacts. They are featured to be either conscious and unconscious.

Damasio conceives the emotion as representations of the relationship between the organism and the external object. And the feeling of an emotion is considered to be based on "having mental images arising from the neural patterns which represent the changes in body and brain that make up an emotion" [54].

According to the interpretation accepted in this thesis, this can be described under computational terms as follows: emotion is the representation of relationships between the system, and the effects that an external object causes on the system. And the feeling of an emotion is argued by providing *c-Images*, arising from the *c-Maps* which represent changes related to the effects caused by the object.

This is the main reason that justifies the requirement of a *Process of Emotional Perception*. From the point of view of Damasio, the emotion phenomena is framed within a structure of four coarse grain processes: (a) basic regulation, (b) emotion, (c) feeling, and (d) higher reasoning about feeling. Some additional actions are thus required beyond those aiming at the categorization of emotion (i.e. the structure by means of which emotion can be categorized with *Emotion Attribute* vectors). Essential features in the model are needed in order to: (a) trigger the essential processes about the basic regulation, (b) create the relationships that characterize the emotion and the feeling (i.e. *c-Maps*, (c) create the images that represent these relationships (*i.e. c-Images*, and (d) allow for the connection with the structures of the self in order to enable processes of high level reasoning.

Thesis 16 ♣ *Justification of MEP*
The *Model of Emotion Perception (MEP* is an artifact aimed to obtain the required relationships (c-Maps) and patterns of states (c-Images) in the form of an inner-object. The objective is to obtain an artifact that integrates all these relationships (c-Maps) in order to make later representations that can be integrated as images-like representations in *AGSys* (i.e. exploitable models).

The following paragraph describes an interpretation of the theory approached by Damasio for the purposes of this thesis. This is one possible approach and it is not argued to be the mirror image of the theory (since those parts without an accurate explanation have been conceived as plausible computational procedures). When Damasio describes his theory, he refers to emotion as the final emotion, an understanding that he shares with James [118]. And he refers to feeling according to the conceptual perception of this emotion, referring thus to the final feeling regarding any life system.

Thus, even is *MEP* uses analogous processes and names to those used by Damasio, this will result in the creation of an object that integrates all the information required for the later interpretation of the experience. And this interpretation is conceived within the work of a different additional process.

> The conceptual difference among *artificial emotion*, *computational emotion* and *emotion-Object* is here highlighted. The former artificial emotion is conceived as the consequence of the process (i.e. the computational emotion). The *emotion-Object* is referring to an artifact used in *MEP* and that constitutes a previous phase to the representation of emotion.

I. General overview

> Summary of the approach by Damasio, with the interpretation assumed for the purposes of this thesis (see Chap. 4, Sect. 4.7):
>
> The problem of feeling is described according to two matters:
>
> 1. How the system creates *c-Images* of an object.
> 2. How the system concurrently might cause "the sense of self in the act of knowledge".
>
> The problem is solved based on three essential assumptions:
>
> 1. By accepting bodily changes as an essential foundation of emotion.
> 2. By postulating a "First-order Bodily Representation": conceived as a causal relationship between first order maps, and the representation in the form of second order maps.
>   2. (a) "First-order maps": c-Maps that represent the 'system' and 'the external object'.
>   2. (b) "Second-order maps": Represent the relationship between the 'system c-Map' and 'the external object c-Map'.
> 3. Postulating a "Second-order Bodily Representation": it is the problem of transforming the "Second-order maps" in c-Images.
>
> Proposed explanation by Damasio: first-order and second-order maps are related to the body, and the c-Images that illustrate this relationship can be considered feelings.
>
> About the emergence of consciousness he proposes:
>
> 1. First component of Consciousness: in which the system is represented by the *proto-self* (♣ 50).
> 2. Second component of Consciousness: which involves the subsequent consequences of the image created by the former.
>
> Regarding the emergence of object representation: it is argued to happen when it is focused on the object (i.e. attention is paid to an object)
>
> Finally, the substrate for feeling emotion is conceived on the basis of two types of changes (♣ 53):
>
> 1. Changes related to physical state.
> 2. Changes related to cognitive state.

As previously introduced, even if the following approach is based on the theory of Damasio, it is not a mirror model of this theory (Fig. 5.39). This approach assumes some essential foundations that should be taken into account:

1. The following approach proposes a method aimed to integrate information, in order to be used later for the purposes of approaching the experience.
2. Names analogous to those proposed in the original theory are used. However, they are not exactly modeling the original element.
3. Three intermediate objects are considered in order to recursively obtain the *feeling-Object*: the *emo-inner-Object*, the *arousal-Object*, and the *emotion-Object*.

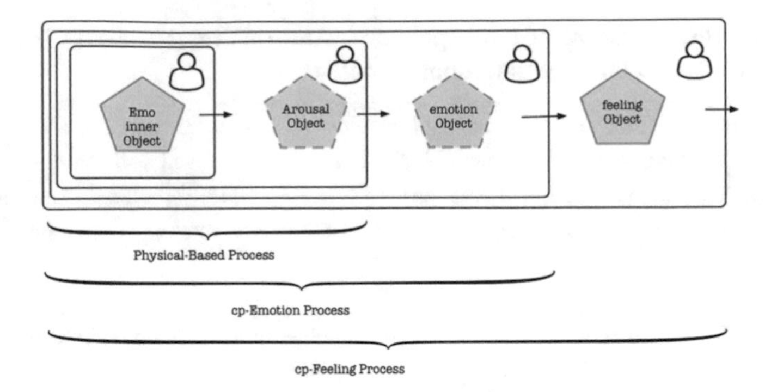

Fig. 5.39 MEP builds the feeling-Object by means of three recursions of R-Loop: **a** The Physical-based process, **b** cp-Emotion process, and **c** cp-Feeling process

4. The *feeling-Object* will represent the object that integrates the full range of information related to the changes that cause the emotional event. The representation of this object will be a matter of additional processes.

The *Model of Emotion Perception (MEP)* is conceived according to four essential processes based on the idea of loop related to the *R-Loop*: the three former processes are aimed to build the *feeling-Object*, and the latter is aimed to approach some type of integration between this *feeling-Object* and the bases that ground the self. Each of these four processes constitutes a unitary loop under the principles of the *R-Loop* approach.

The processes are: (a) the physical-based process that motivate the construction of the *arousal-Object* from the source of the *emo-inner-Object*, (b) the *cp-Emotion* process that takes this *arousal-Object* in order to build the *emotion-Object*, and (c) the *cp-Feeling* that takes the *emotion-Object* in order to construct the *feeling-Object*.

Finally, there is a final loop in order to approach some type of integration between this *feeling-Object* and the bases that ground the self. And this is related with the previously described problem of representation.

II. The Physical-based process Under the principles of the *R-Loop*, it is required to identify: (a) the identity of the object that affects this stage, and (b) the effects caused on *AGSys* that can be used to infer and trigger actions of basic regulation (i.e. action readiness). These two entities will define the *First-order maps*, and the relationship between them will define the *Second-order map* (see Fig. 5.40).

The *Universe of emotions* provides the grounds to obtain these objects. This phase is conceived as related to the performance of the *c-emoGoals*, and it affects the behavior of each unitary *c-emoGoal*, as well as the created environment in which they interact (with the objective of analyzing this interaction).

The *Universe of Emotion* can be analyzed thus by means of the state of the *appraisal dimensions (a-Dimensions)*. These *appraisal dimensions* can be evaluated by means of the analysis of *c-Concerns*, and each *c-Concern* by means of the analysis of *c-emoGoals*. We highlight the requirement of a dynamics analysis, in

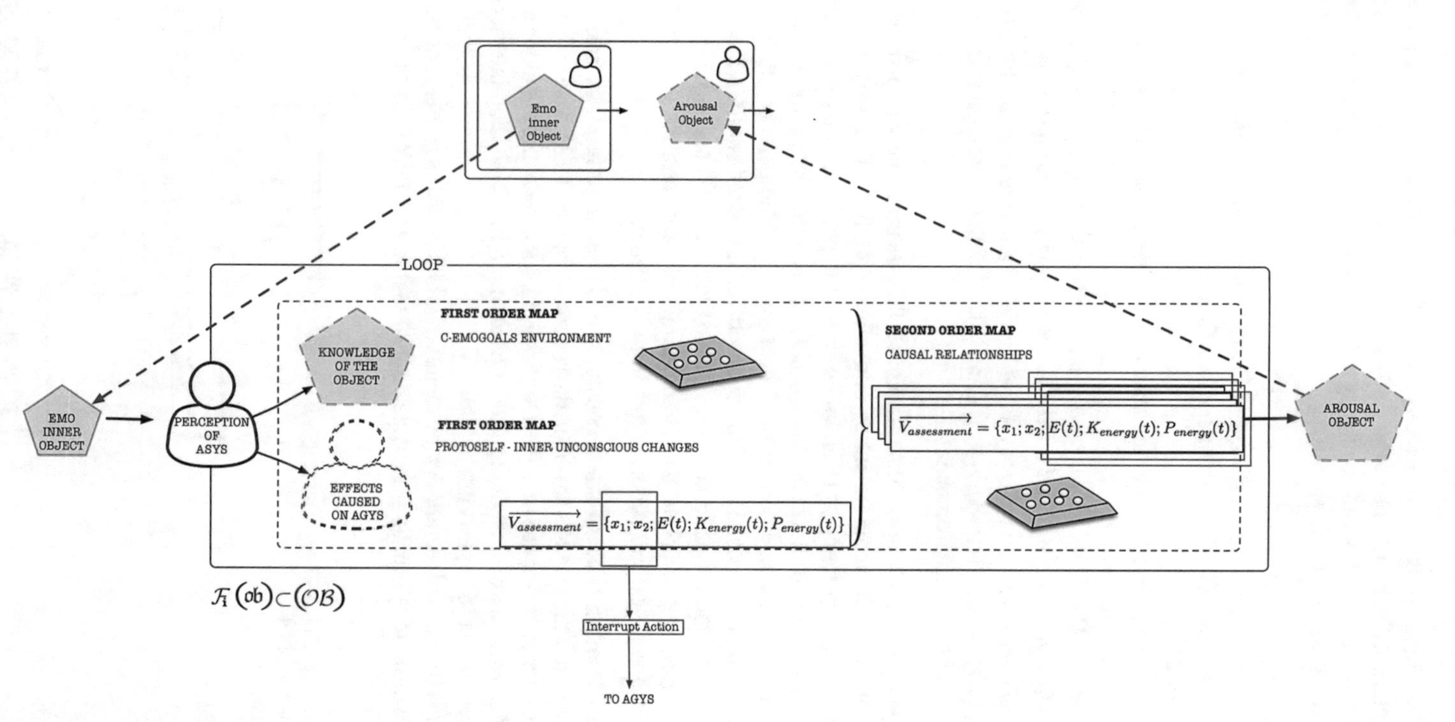

Fig. 5.40 Conceptualization of the first stage in MEP

order to conceptualize the unconscious *proto-self*. The proposed model allows for triggering alarm-calls by means of the analysis of the state variables related to each *c-emoGoal*: x_1 and x_2. This enables the system to trigger basic regulation related to the *c-emoGoals* (see rationale ♣ 50).

1. *First-order maps*
   (a) the environment created for *c-emoGoals* interaction, and (b) the set of state variables (x_1 and x_2) related to each *c-emoGoal*.
2. *Basic regulation*
   The series of triggers related to the state of the state-variables in *c-emoGoals*.
3. *Second order-order maps*
   They represent the relationships among the First-order maps. Some of these relationships might be considered according to (a) the general question about the feature of energy related to each c-emoGoal state (i.e. the state of $\overrightarrow{V_{assessment}} = \{x_1; x_2; E(t); K_{energy}(t); P_{energy}(t)\}$), and (b) additional assessments on the basis of the state of their environment.

In computer science, *interruptions* are signals that are triggered within the systems in order to indicate the requirement of immediate attention. This way, processors are alerted to a high priority state in which they stop processes that are defined as non critical, to attend to those that are considered as more important according to the priority that causes the interruption. This basic regulation can be performed by means of a finite state machine (FSM) (i.e. a behavior based runtime model) that triggers calls for interruption to *AGSys* related to the state of (x_1 and x_2). Figure 5.41 illustrates a generic conceptual model as a graph of instances. As a result of this stage, the *emo-inner-Object* becomes the *arousal-Object*.

III. The cp-Emotion process Analogously, as in the previous loop, the *Universe of emotions* provides the grounds to obtain the required objects. This phase is conceived as related to the performance of the *c-Concerns*, and it affects the behavior of each unitary *c-Concern*, as well as the created environment in which they interact (with the objective of analyzing this interaction).

Arousal is the state of excitement by means of which some essential processes are triggered in order to optimally face a punctual situation, and it constitutes the ground

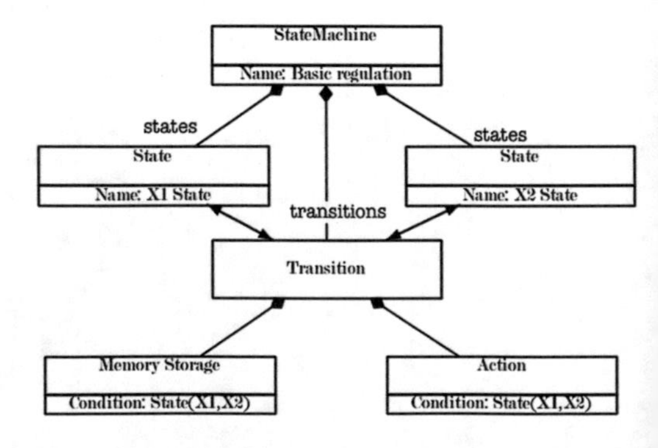

Fig. 5.41 Conceptual model of Basic regulation as a graph of instances

for emotion [5, 54, 107, 118]. We will conceive these arousal states in artificial systems as interruption states, that is, a state in which alerts notify the system about the requirement of immediate attention regarding some source that is establishing high-priority (see rationale ♣ 51).

Right now, the initial object of this stage is the *arousal-Object*. It has been conceived as belonging to this stage and not to the previous one because it is now that it integrates the information about the basic processes of regulation. Since this is an essential feature of *Arousal*, it has to be integrated within the *arousal-Object*.

The process of growing the object is analogous to the previous phase (under the principles of the *R-Loop* operation). This time, however, the dynamics are related to the dynamic state of *c-Concerns* instead of the dynamic state of the *c-emoGoals*. Hence, as illustrated in Fig. 5.42:

1. *First-order maps*
   (a) the environment created for *c-Concerns* interaction, and (b) the set of state variables (x_1 and x_2) related to each *c-Concerns*.
2. *Basic regulation*
   The series of triggers related to the state of the state-variables in *c-Concerns*.
3. *Second order-order maps*
   They represent the relationships among the First-order maps. Some of these relationships might be considered according to (a) the general question about the feature of energy related to each *c-Concerns* state (i.e. the state of $\overrightarrow{V_{assessment}} = \{x_1; x_2; E(t); K_{energy}(t); P_{energy}(t)\}$), and (b) additional assessments on the basis of the state of their environment.

This basic regulation can be conceived again in the form of a finite state machine (FSM) that triggers calls for interruption to *AGSys* related to the state of (x_1 and x_2). The same Fig. 5.41 illustrates this generic model, but taking into account that state-variables are related to the *c-Concerns*. As a result of this stage, the *arousal-Object* becomes the *emotion-Object*.

IV. The cp-Feeling process Also in a similar way, the *Universe of emotions* provides the grounds to obtain the required objects. This phase is conceived as related to the performance of the *a-Dimensions*, and it affects the behavior of each unitary *a-Dimension*, as well as the created environment in which they interact (with the objective of analyzing this interaction).

Right now, the initial object of this stage is the *emotion-Object*. It has been conceived as belonging to this stage, and not to the previous one, because it is now that it integrates the information about the arousal (situational state of concerns, goals and basic regulation). Since this is an essential argument of Damasio (in accordance with the theory of James [118]), it requires to be integrated within the *emotion-Object*. So that, as illustrated in Fig. 5.43:

1. *First-order maps*
   (a) the environment created for the agent *a-Dimension*, and (b) the set of state variables (x_1 and x_2) related to this *a-Dimension*.

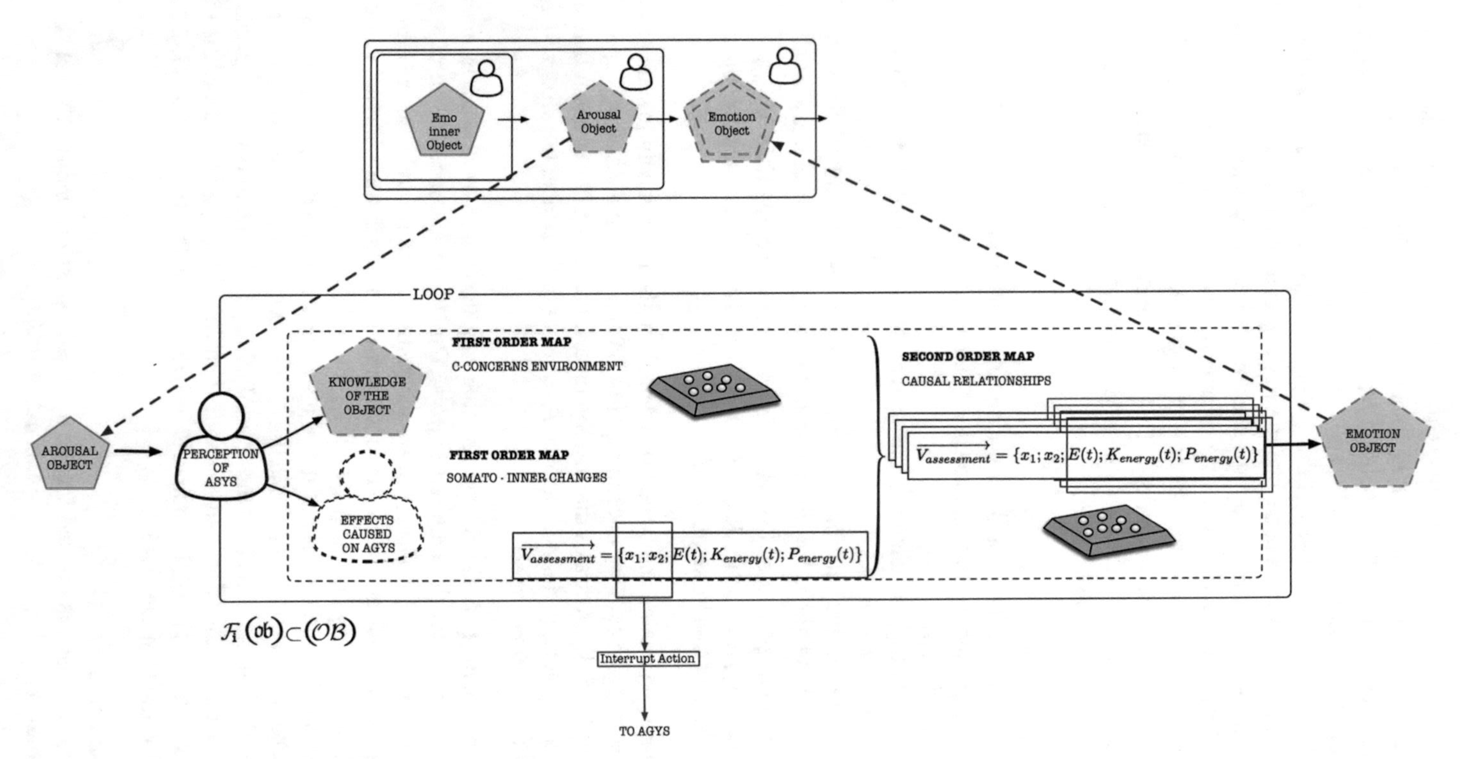

Fig. 5.42 Conceptualization of the cp-Emotion stage in MEP

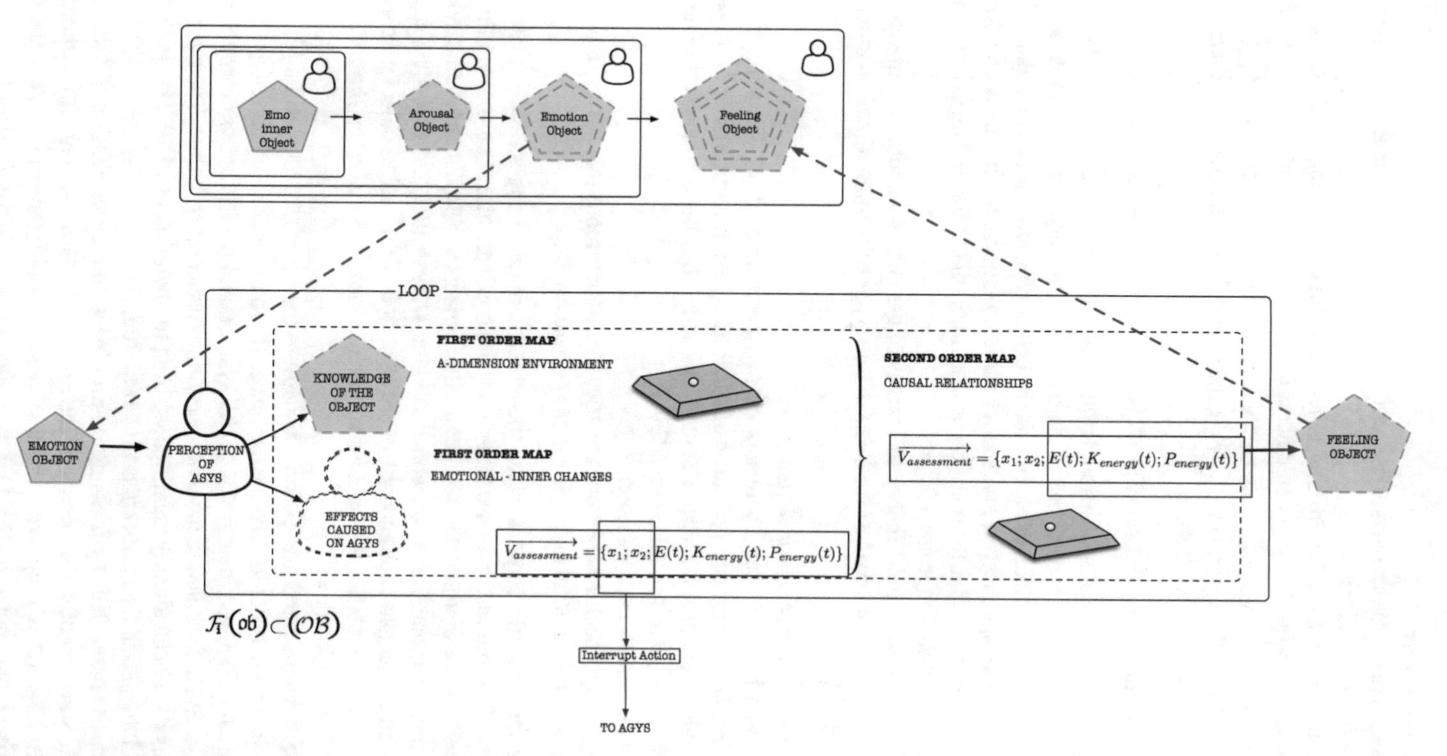

Fig. 5.43 Conceptualization of the cp-Feeling stage in MEP

2. *Basic regulation*
   The series of triggers related to the state of the state-variables in this *a-Dimension*.
3. *Second order–order maps*
   They represent the relationships among the First-order maps. Some of these relationships might be considered according to (a) the general question about the feature of energy related to this *a-Dimension* state (i.e. the state of $\overrightarrow{V_{assessment}} = \{x_1; x_2; E(t); K_{energy}(t); P_{energy}(t)\}$) and (b) additional assessments on the basis of the state of its environment.

As a result of this stage, the *emotion-Object* becomes the *feeling-Object*.

V. The approach to representation Right now, the initial object of this stage is the *feeling-Object*. It has been conceived as living in this stage and not in the previous one, because it is now that it integrates the information about the emotion (situational state of dimensions, concerns, goals and basic regulation). Since this is an essential argument of Damasio (i.e. the emotion as the initial object for feeling), it requires to be integrated within the *feeling-Object*.

Finally, according to the theory of Damasio, the analogy about the substrate for feeling emotion is conceived on the basis of two types of computational changes (see rationale ♣ 53):

1. *Changes related to physical state*
   They will be conceived by means of two mechanisms: (a) The creation of a *Physical Loop* within the operation of MEP, and (b) the creation of an *Anticipative loop* which enables optimized behaviors (which has been conceived under the operation of state-variables and a FSM)
2. *Changes related to cognitive state*
   They prompt specific behaviors, changes in the continuous processing of physical states and modulation in the processes of assessment.

After analyzing the physical state, the analysis of the changes related to the cognitive state is required. There are three essential concepts that should maintain their individuality to be distinguished from all other objects within the emotional system. That is, three computational concepts that should have entity and should be susceptible of being changed continuously, over the period of time of an object perception. They are the computational analogies of the *somatosensory system*, *proto-self* and *core-self* (see rationale ♣ 39).

These three essential components constitute the basic elements to build: (a) the *first component* in order to generate a *c-Image* like a non-verbal relationship between the external object and the physics of the system, and (b) the *second component* in order to generate the consequences of the *c-Image* concerning feeling. The two former analogies have been already identified within the operation of MEP. The analysis of how to characterize the *core-self* is now required.

Damasio argues that, by focusing on an object, the representation of that object emerges. Subsequently, the arising of the *core-self* (conscious but still without reference to *identity*) is proposed. And it is when autobiographical records are made explicit in reconstructed images that they become *autobiographical-self*.

This memory has been previously described in this chapter (see Fig. 5.36). It is assumed by Damasio as the one architecturally connected to the *core-self* and the *proto-self* at each time (i.e. a lived instant). And *consciousness* defines the integration of the *core-self* and *autobiographical-self*.

We need to conceptualize the *core-self* as an artifact that will engage the focused *attention on an object* required in this process. We argue that this attention emerges from the act of assessing something that is relevant. As we have focused on emotional relevances, this can be assumed as the act of assessing something that is emotionally relevant for the system. And it comes from the basis of a logical structure of relevances that the system deploys and uses to assess (see rationale ♣ 49).

Thesis 17 ♣ *The problem of Autonomy in AGSys*
The state of the *Universe of Emotions* will be considered in this work as the foundation of this representation. That is, the *core-self* will be considered according to the conceptualization of the situational state of the *Universe of Emotions*. This way, the requirement of 'attention on an object' is integrated: the situational state of the *Universe of Emotion* changes just as a result of relevances within the state of *c-emoGoals*, *c-Concerns*, *a-Dimensions* and their interactions. This universe does not change due to direct errors within the *inner-Environment* of *AGSys*. These errors are transferred to the environment of an agent-based model that behaves in accordance to them. And it is just as a result of a relevant behavior that the situational state of this universe becomes relevant: *c-Concerns* start to move and it is in this way that the 'attention on an object' can be assumed. ♣ Consequently, the situational state of the *Universe of Emotion* ca be considered to be the artifact *core-self*: *AGSys* is aware about this state without reference to any *AGSys identity*.

This is in accordance with the description of how representation can be conceived regarding *AGSys*. The *Emotion Attributes* are used to represent the situational state of the *Universe of Emotion* as well as to, afterwards, categorize the own emotional state (i.e. basic emotions such as those of Ekman [67]). They are also assumed as the environment monitored by the loop (i.e. the adaptive engine) related to the autobiographical knowledge (see Fig. 5.36).

This way, an additional recursion (within the operation of R-Loop) can be conceived in order to approach some type of ground for experience in *AGSys* (see Fig. 5.44):

1. *First-order maps*
   (a) the conceptualized environment of the *Universe of Emotion*, and (b) the related state of the *Self-Memory-System*.
2. *Basic regulation*
   The series of triggers related to the state of the *Working-Self Goals* concerning discrepancies between the *Working-Self Domains*.
3. *Second order-order maps*
   The relationships among the First-order maps are represented by means of the operation of the *Self-Memory-System*. Some of these relationships might be considered according to the work of the adaptive engine related to the *Working-Self*

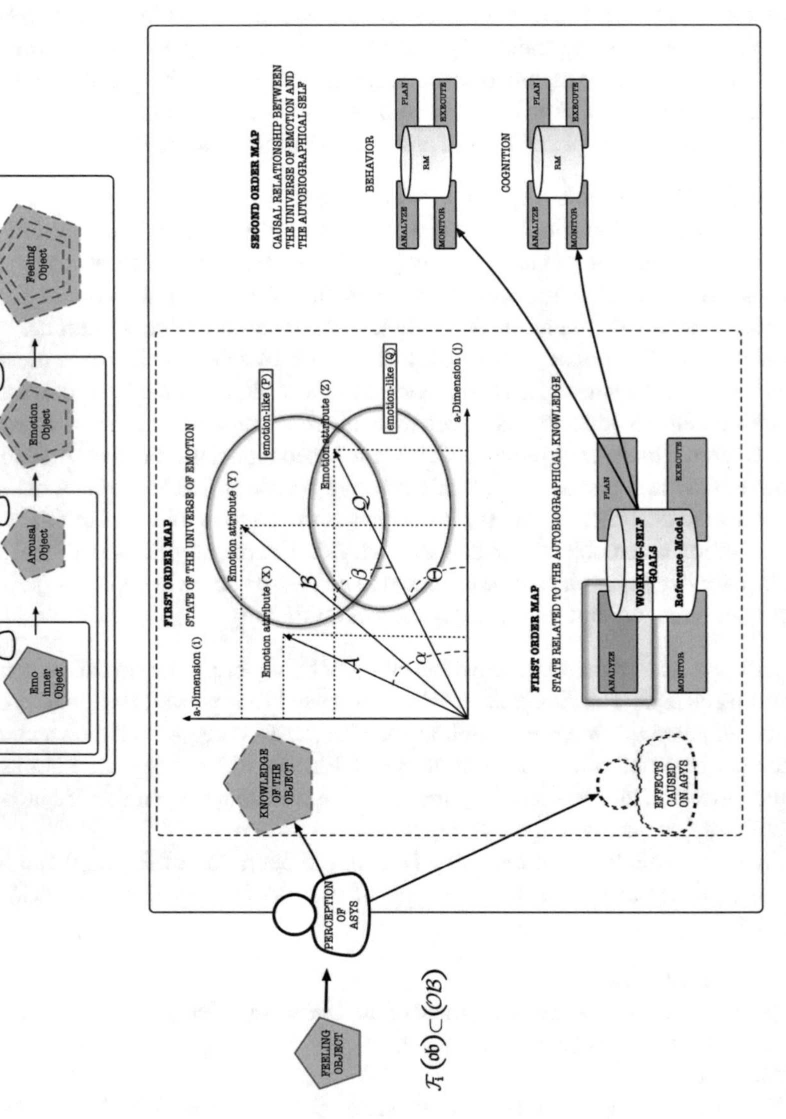

Fig. 5.44 Conceptualization of the representational stage in MEP

Goals, which execute changes in behavior and cognition in order to solve discrepancies among these goals. This way, the high-level reasoning model uses knowledge about emotional goals stability, hence allowing for new directions in which mission goals might be assessed according to the situational state of this stability.

Working-Self Goals are conceived in this thesis as the model that reflects the adaptable system and its environment. This adaptable system is the amount of *Emotion Attributes* and the environment, the *Universe of Emotion*.

As a result of this stage, the *feeling-Object* can be related to the *Working-Self Goals*. And somehow, this relationship together with the imagery created along the entire process of MEP, might become an approach to the experience of feeling.

VI. Considerations about imagery All the previous essential components (i.e. protoself, somatosensory and core-self conceptualizations) constitute the basic elements to build: (a) the *first component* in order to generate a *c-Image* like a non-verbal relationship between the external object and the physics of the system, and (b) the *second component* in order to generate the consequences of the *c-Image* concerning feeling.

According to the approach herein described, the *c-Map* can be considered as the entire object obtained from the basis of the operation performed by *R-Loop* and *MEP* (with the conceptualization of *c-Map* assumed in this thesis).

1. *c-Map*
   It is assumed as the structure of data obtained within the work of MEP.
2. *c-Images*
   They are conceived as the way in which *AGSys* represents this structure of data (i.e. c-Map). And it can be obtained from the already described work of ontological artifacts. This way, by means of an inner ontology, *AGSys* might obtain exploitable knowledge related to the obtained *c-Map*. And this can be considered as *conscious c-Images* for *AGSys* (see Fig. 5.45).

Consequently, *AGSys* is able to describe the same reality (i.e. the external object) by means of two different representations: the representation that refers to the logical reasoning (i.e. the *inner-Object*), and the one that refers to the conclusion reached on the basis of the emotional assessment (i.e. the associated *c-Images* that result from the original *emo-inner-Object*).

Emotions are system-wide signals driving the pattern-based reconfiguration process, and when represented at the level of awareness they constitute the emotional feelings. The pattern of data *c-Map* showed in Fig. 5.45, might represent this pattern in order to face the design of reconfiguration processes.

In our theory we consider that the conceptualized situation is the functional coupling of the agent with its environment. This proposed pattern of data represents the coupling of the effects that an external object causes in our agent (i.e. the coupling of *AGSys* and the environment). Hence, the functional design on the basis of this pattern might be the artifact in order to build this functional coupling of the agent with the environment.

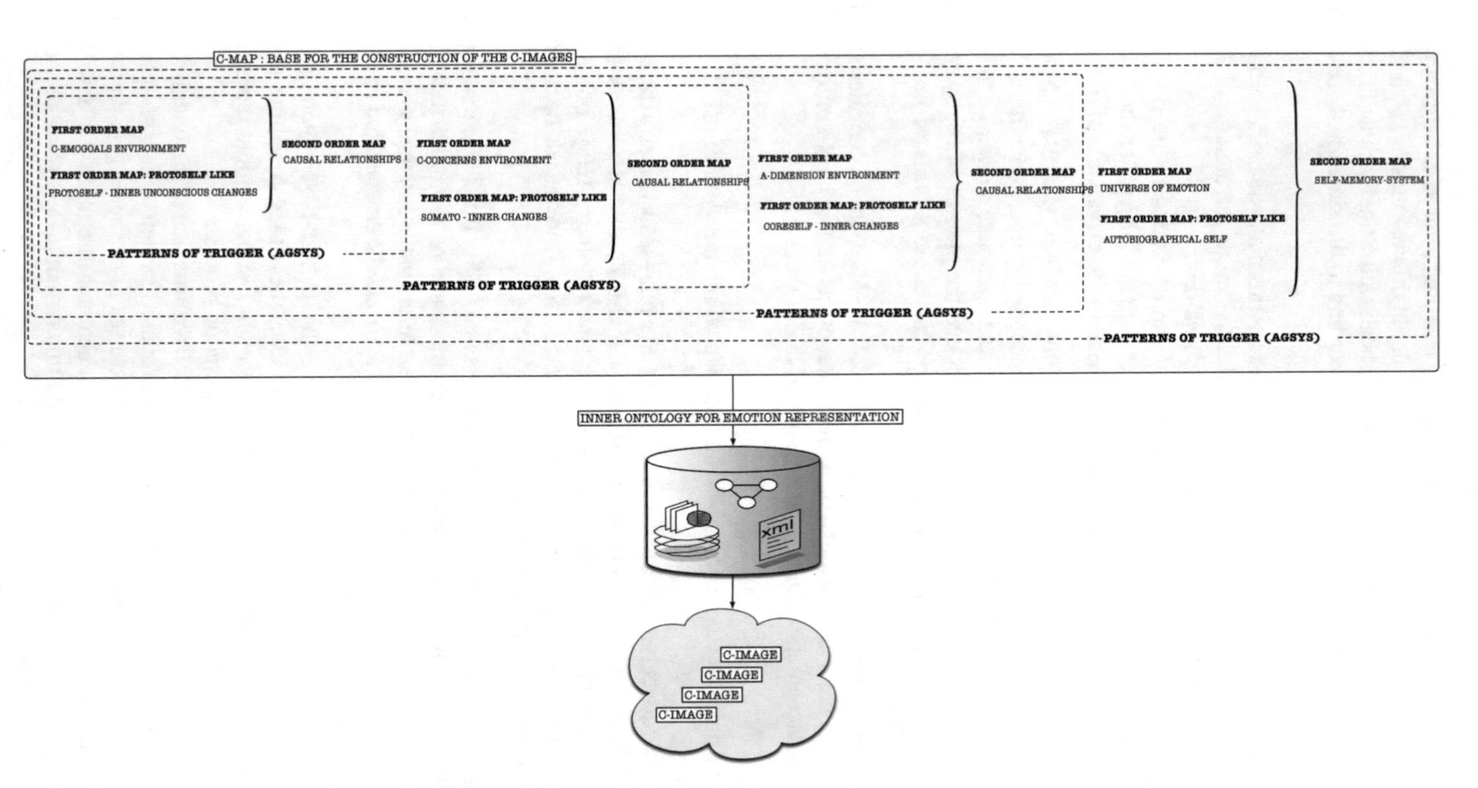

Fig. 5.45 General view of the proposed solution aimed to approach the emotion representation

As we have argued in previous works, emotions are hence conceptually related to the functional structure of the agent. The set of concepts behind them provide an ontology that constitutes the semantic substrate of the morphofunctional elements of the agent (see Sanz et al. [231]). And it is finally obtained in this thesis by means of the *c-Image* artifact.

This way, *AGSys* is able to describe the same reality with two different representations (see Fig. 5.46): the representation that refers to the ontological meaning (i.e. the *inner-Object*), and the one that refers to the emotional assessment (i.e. the associated *c-Images*) that result from the original *emo-inner-Object*. And the emotional ontology will be defined on the basis of the complete amount of obtained information.

5.8 Architectural Foundations

We have described a theoretical approach in order to support the entire process of an emotional assessment related to an external object. The work of computing value is carried out by *ESys*, which performs the emotional operation on the basis of transversal data concerning *AGSys*. These transversal data are conceived to be obtained by means of reflex agents which are monitoring specific parts of *AGSys*, being these parts defined according to emotional requirements.

We will assume that reasoning is a matter of *AGSys*, and that *ESys* builds new information about the transversal state which is sent to *AGSys*.

It is thus required to establish the basis for developing a logical model of interaction between *AGSys* and *ESys*, since their reciprocal action and influence is an essential need for the particular purpose of emotion performance.

Since *AGys* is a model-based cognitive agent, a cognitive approach of emotion appraisal has been taken. It is then accepted that emotions are constructed on the basis of emotion-based assessments. By analyzing the term *"assessment"*, it seems to be related to the idea of reasoning. When we talk about an emotional assessment, it directly implies the existence of emotional references.

Once analyzed the assessment itself, the question now is how this assessment is structured within the environment of interaction *AGSys–ESys*. A general view of this interaction is illustrated in Fig. 5.47. However, since the functional architecture of the reciprocal actions will not be formalized, additional explanations are required in order to clarify essential features related to their interaction.

5.8.1 The Environments of AGSys

We will assume that *AGSys* is the one that interacts with the external environment. The work of *ESys* is performed within the *inner-Environment*, the one created to transversally monitor the *AGSys* performance. This virtual environment is supported

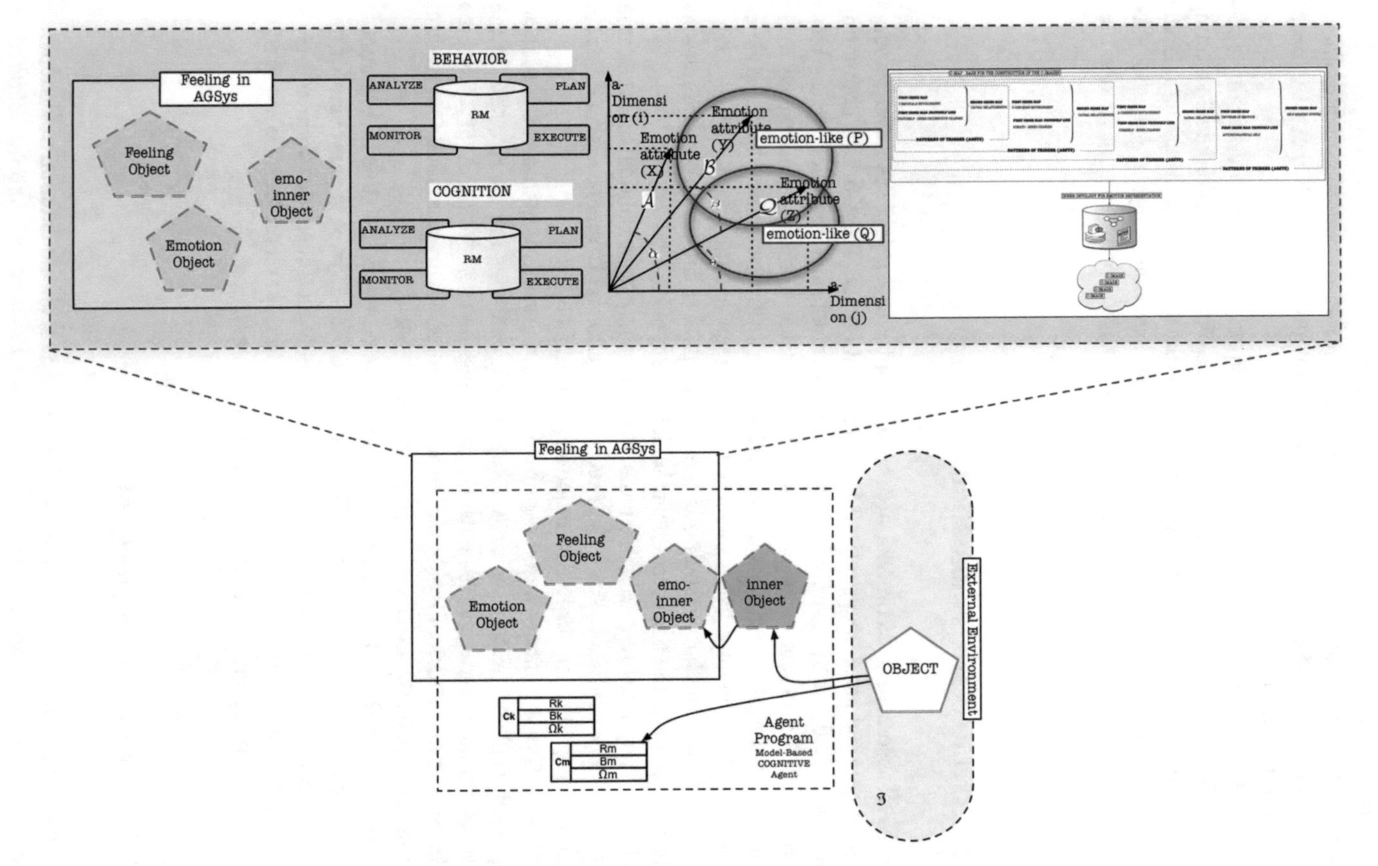

Fig. 5.46 General view of two representations made on the basis of the same external object

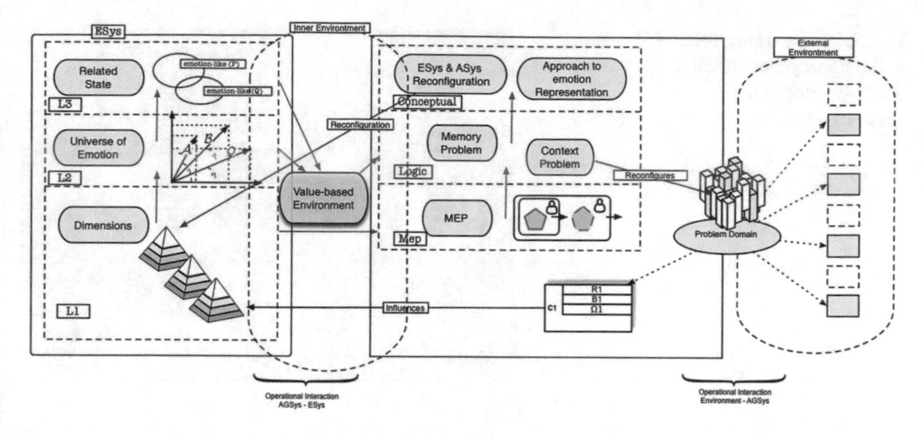

Fig. 5.47 General view of AGSys–ESys interactions

by the work of different reflex agents that are monitoring a required amount of *appraised-Subsystems* defined under emotion based requirements.

This *inner-Environment* thus provides data from *AGSys* to *ESys*. However, we assume that once *ESys* computes the state of the emotional structure of goals (i.e. information related to the *Universe of Emotion*, the *Emotion Attributes* and the emotion-like state), this knowledge is required in order to obtain the *'approach to the experience*. This part of work is performed by *AGSys*, so that *AGSys* needs additional data from *ESys*.

Hence, the existence of an additional inner environment that might be taken into account during the phases of implementation is assumed. *AGSys* builds the *feeling-Object* on the proposed basis of the agent-based system of goals. But the approach to the experience is built on the basis of those values that regard the *Emotional Attributes* and the *emotion-like state*.

Consequently, we consider the conceptualization of a *Value-based environment* embedded within the *inner-Environment*, which provides *AGSys* with those values that are required in order to build an approach to the experience of emotion (see Fig. 5.48).

5.8.2 AGSys–ESys Reciprocal Actions

The concept of abstraction is commonly used in computer science in order to draw levels of complexity to computationally deal with every level separately. The intricate structure of emotional goals herein proposed involves factors, processes, and environments that are better managed by considering their operations separately. The work of *ESys* will thus be considered according to three levels of work: (a) the work related to the emotional objectives (i.e. *c-emoGoals*, *c-Concerns*, and *a-Dimensions*),

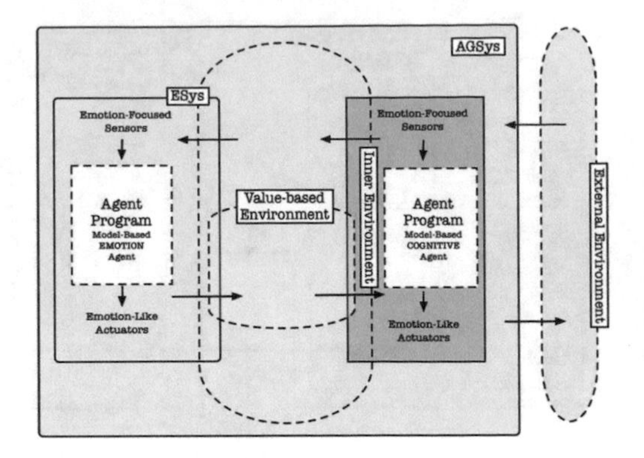

Fig. 5.48 Overview related to the conceptualization of the three proposed environments

(b) the work related to monitor the situational state of these emotional objectives (i.e. the mutual influence among them), and (c) the emotional state regarding the monitored situational state. This way, the emotional assessment (*ESys*) and the reasoning about the emotional assessment (*AGSys*) are improved (see rationale ♣ 55).

Analogously, *AGSys* will be conceived according to: (a) the work related to the construction of the *feeling-Object*—which concurrently allows for basic regulation, (b) the work related to the references between this *feeling-Object* and how it affects the working-self goals and autobiographical knowledge, and (c) the work related to the ontology that provides *AGSys* with semantics about the whole state (see Fig. 5.49).

5.8.3 Essential Questions Regarding the Appraisal

This approach assumes a series of steps in order to capture the essential elements that build an emotion. Once the error is detected, the environment of emotional goals starts its movement and it implies changes in their environment. Those changes in *c-emoGoals* are sent to *AGSys* and the first recursion in *MEP* starts. However, whether processes of basic regulation related to these goals are triggered or not, it is just when *c-Concerns* move from their state that this process becomes conscious for *AGSys* (see Fig. 5.50).

This is in accordance with the theories of appraisal work supported in previous sections: analogously to the theory of Ellsworth and Scherer, there is no assessment of two *c-Concerns* at the same time (see rationale ♣ 17 in Chap. 4, and Fig. 5.21 in this chapter).

Paying particular attention to this feature, it is essential to clarify how this issue is conceived in this thesis. Computationally speaking, this is a property by means of which an amount of computations are being simultaneously executed. This *MEP* process is an extremely costly process regarding system resources. Nowadays, systems usually employ several processors for the computations and concurrency is used for higher levels of activity. However, *MEP* is built in order to obtain the coupling

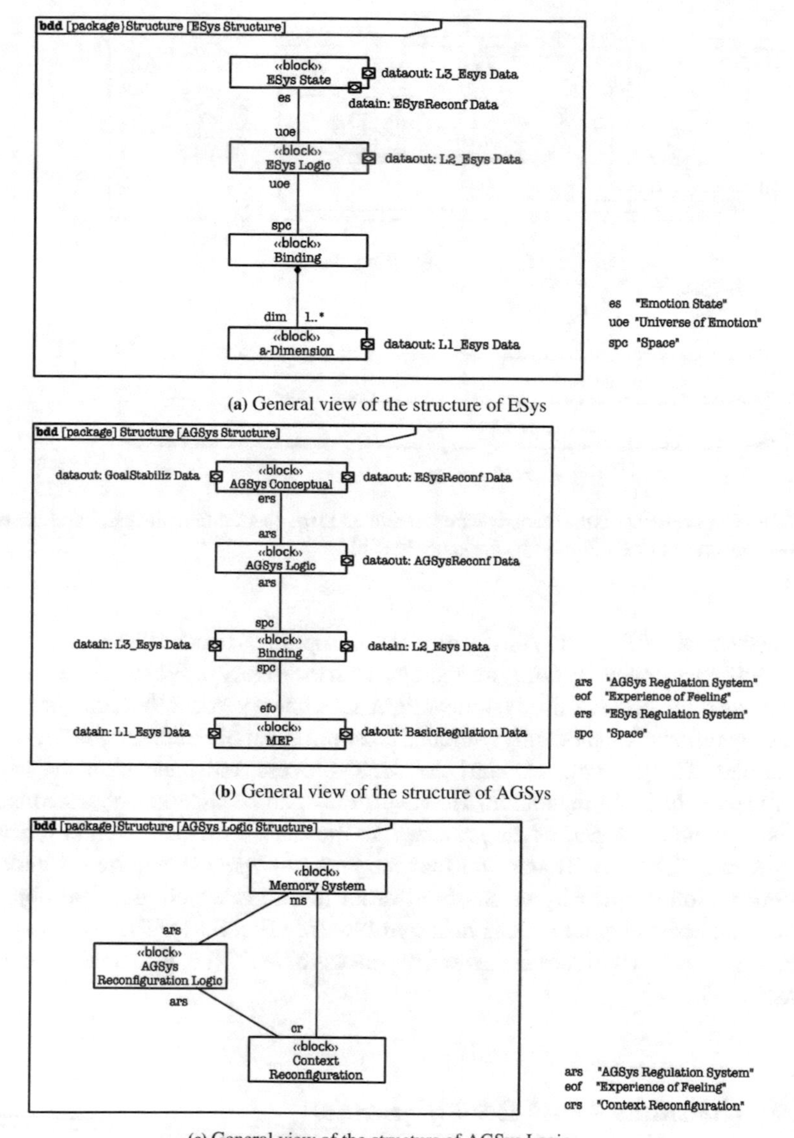

(a) General view of the structure of ESys

(b) General view of the structure of AGSys

(c) General view of the structure of AGSys Logic

Fig. 5.49 General view of the structure of and the reciprocal interaction between AGSys and ESys. AGSys logic has been included in order to illustrate the interaction with the context reconfiguration and the memory system

of *AGSys* with its environment related to a concrete object. Hence, according to the perceptual foundation related to recursion, it cannot be performed within different processes. Consequently, it cannot be distributively performed in different processors.

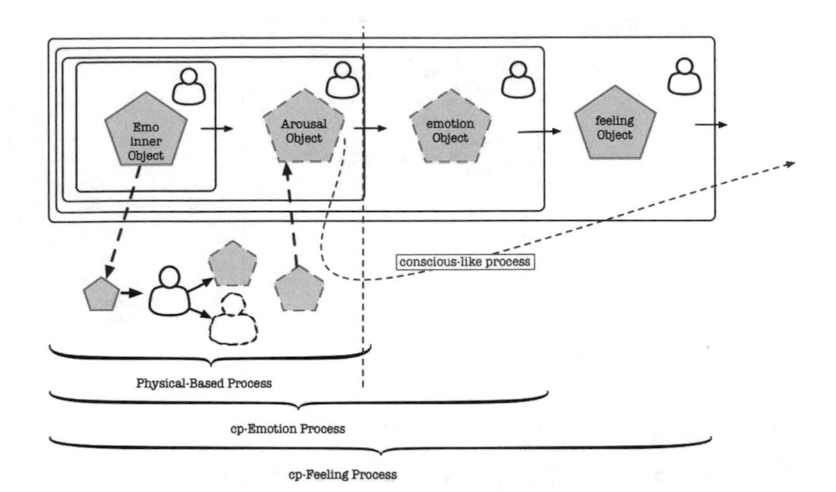

Fig. 5.50 Once MEP starts the second recursion related to the cp-Emotion process, the concurrency of processes related to additional objects is not feasible

However, as *MEP* is a very costly process, the option is to coordinate the execution and avoid concurrency. By doing an analogy with the theory of Ellsworth and Scherer, it seems as if the brain of life systems might save energy under the same principles.

The deviation of *c-emoGoals* without the requirement of building a *feeling-Object* is assumed. This *c-emoGoals* start the *MEP* process and they might be triggering processes of basic regulation. However, they can be assumed as no-conscious processes which take part of the *protoself* in the sense of Damasio. It is just when changes in *c-Concerns* are relevant that the *feeling-Object* is required. Hence, the implementation of intensity detectors in order to assess which 'external object' is the one that causes higher actions inside *AGSys* (see Fig. 5.51). The *c-Concern* that raises higher values will start the recursive process of *MEP* (i.e. the attention to which Damasio referred).

5.8.4 Questions About Reconfiguration

We agree that emotion reconfigures the cognitive system in order to return the system to the equilibrium. That is, there is an action from the *ESys* to *AGSys* in order to obtain a method for *reasoning–concerning–value*. Additionally, a mutual influence between *AGSys* and *ESys* will be assumed in the sense that: *ESys* is monitoring the transversal state of *AGSys* and influences the control of *AGSys* on the basis of its work. And *AGSys* should have some negative feedback in order to reduce the influence of *ESys* under required circumstances (see rationales ♣ 57 and 59).

In general, reconfiguration is driven by processes that try to match the organization of the agent to what is needed to accomplish a certain goal. In essence, they are

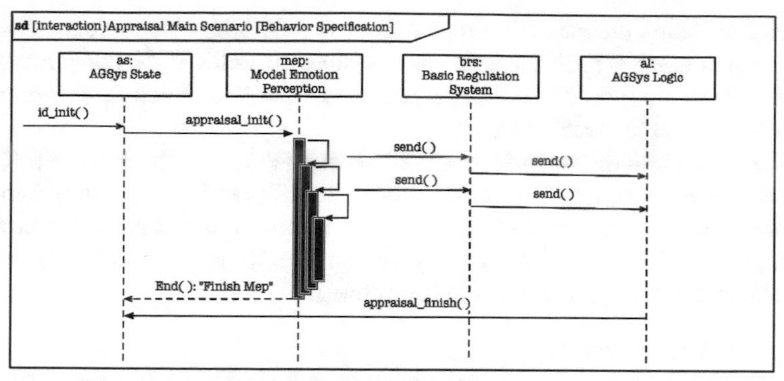

(a) Appraisal work

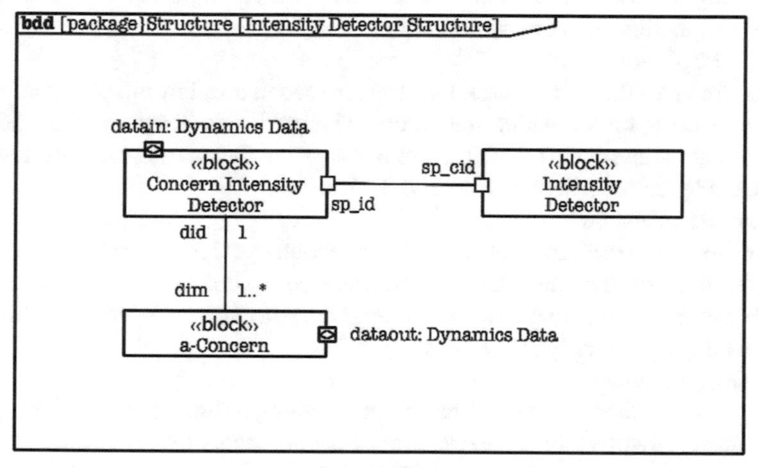

(b) Intensity detector

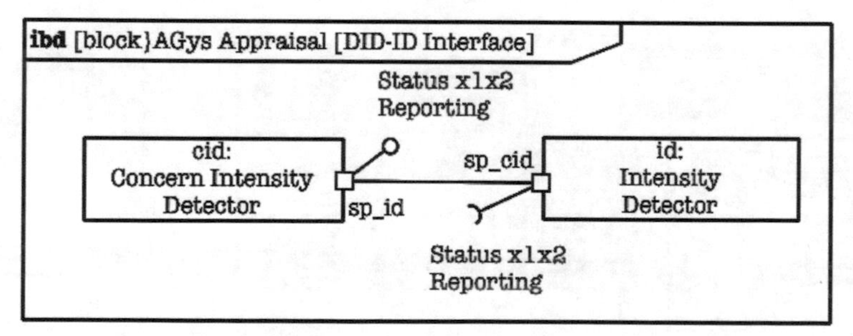

(c) Interface between intensity and concerns detectors.

Fig. 5.51 Figure **a** shows a general view of the appraisal work related to a concrete object. Figure **b** describes a possible essential structure of the intensity detector that obtains the higher values of $x_1 and x_2$. And figure **c** describes the interfaces between these two detectors showing a connector that joins this property via standard ports

structural feedback processes in order to functionally adapt the system to its environmental circumstances. This thesis has conceptualized the *Universe of Emotion* in order to obtain a region of the *AGSys* emotional space in which to trigger control actions performed by controllers.

This reconfiguration however needs to be assumed from different perspectives and, consequently, the existence of different types of functional reconfiguration has to be assumed. It is a matter of future works to design the complete strategy for reconfiguration. However, some essential reconfigurations required for the work of an emotional strategy (Fig. 5.52) can be outlined:

1. *Fast adaptation strategies*
   They refer to those processes that trigger reconfiguration in functional strategies focusing on fast *AGSys* adaptiveness. These are strategies of action readiness, basic regulation and context awareness.
2. *Slow adaptation strategies*
   They refer to those processes that trigger reconfiguration in functional strategies focusing on regulating the emotional experience. They are strategies that reconfigure the way in which the same changes in the environment are assessed, depending on the situational state of *AGSys*.
3. *Learning strategies*
   They refer to those processes that trigger reconfiguration in functional strategies on the basis of previous emotional experiences. To this end, when an emotional experience occurs, it has to be stored together with those elements that allow for matching and memory recovering.
4. *Tuning strategies*
   They refer to those processes that trigger reconfiguration in order to change the parameters involved in the assessment of the emotional experience.

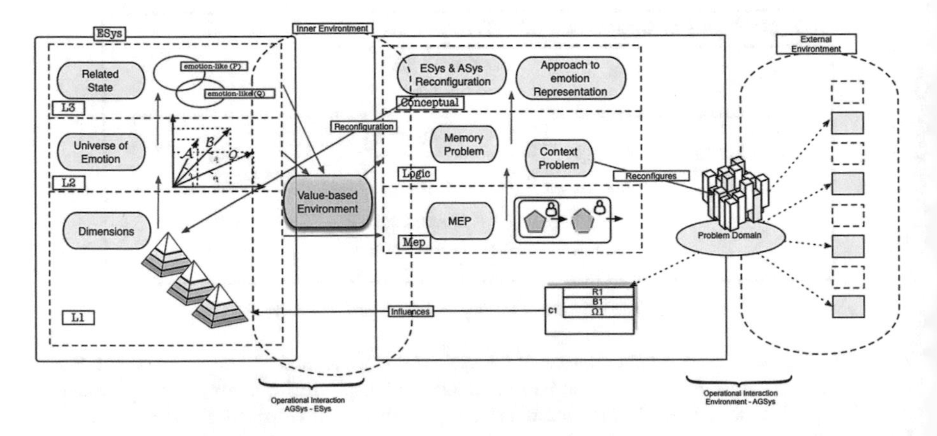

Fig. 5.52 General overview of the reconfiguration strategies in *AGSys* and *ESys*

The proposed approach of this thesis allows for the whole set of reconfiguration types. The structure of the emotional goals allows for joining the external environment with a context-model that is linked to the emotional goal (see Fig. 5.9). It allows for the reconfiguration of the problem domain regarding the external environment when required, under the circumstances of an emotional event. This way, *AGSys* reconfigures resources in order to improve the perception of the environment.

Similarly, the conceived fourfold-role of goals allows to link each goal with its requirements regarding basic regulation when it is required (see Fig. 5.28). I also allows for relating the goal with those chunks of memory that were linked to it in the past, because of an emotional experience. If a system of memories correlation is enabled, it might perform matching directly from perception, and might trigger emotional events before the emotional assessment takes place in *AGSys*.

The behavior of an emotional goal has been characterized concerning an error by means of the design of the evolution of the effects on *AGSys*, rather than directly designing the effects. This way, the behavior of a goal has been characterized so that it is possible to dynamically represent trajectories of changes and stability in every goal. Therefore, the goal's behavior concerning an error has been conceived according to a differential equation as Fig. 5.30 illustrates (a single differential equation has been chosen for the explanation). This way, the behavior of this goal is assessed according to the equation $(AGSys)\ddot{y} + c\dot{y} + ky = error(t)$. And this behavior will differ regarding the values of c and k.

The characterization of the effect of an error changes depending on these values, so that it allows for the reconfiguration of the model's attributes in order to obtain different *emotional tendencies* concerning the same error value.

The fourth role of goals that is associated to the emergence of new requirements is highlighted here. As the goal's behavior has been conceptualized in the form of a differential equation, by using the values of $E(t)$, $P_{energy}(t)$, and $K_{energy}(t)$, we can represent additional values that allow for some type of "emergent system requirements of wellness recovery". This way, we can use these values to design new systems for the emergence of new goals, a relevant feature still unsolved in artificial systems.

Self-adaptation copes with requirements that change at runtime. Therefore, methods are needed so that requirements contribute themselves to be dynamically observed during the execution. Reflection regarding this emergence of requirements can be assumed depending on the value that $E(t)$, $P_{energy}(t)$, and $K_{energy}(t)$ might take at runtime. This way *AGSys* might be able to observe its own goals behavior.

However, this emergence of new requirements should not be understood directly from the value of these features in *c-emoGoals* (i.e. $E(t)$, $P_{energy}(t)$, and $K_{energy}(t)$). They are part of the whole set of needs for the awareness of new requirements, but they are not the only feature. According to Downey, the emergence of a new property is "a characteristic of a system that results from the interaction of its components, not from their properties" [65]. This way, the analysis of the agent-based environment of emotional goals is required together with the analysis of the evolution of these features.

5.8.5 About Agency

Ellsworth and Scherer share a similar perspective regarding some other appraisal theorists about the existence of an appraisal dimension aimed to endow determination, causation or responsibility to the system: *causal-agency* [71]. Even if an accurate description about this *causal-agency* is not needed now, it is however accepted as an influence for the appraisal processes, and linked to the general notion of controllability or coping ability. However, this operation will be assumed within this thesis according to the work of those references related to the autobiographical knowledge and the working-self goals. Since they affect the processes of cognition and behavior, the capabilities of agency and copying will be assumed as part of the work performed by these processes.

However, we remark and strongly assume that this is a simplification made for the purposes of this thesis, and that they are part of the future works in emotion and artificial systems. We agree that emotion reconfigures the cognitive system in order to return the system to the equilibrium. That is, there is an action from the *ESys* to *AGSys* in order to obtain a method for *reasoning–concerning–value* (see rationale ♣ 57).

This way, the higher order reasoning is assumed on the basis of a process that monitors the stability of the *Working-Self goals*. The emotion-based situational state of *AGSys* is monitored by means of *Emotion Attributes*. But the behavior is affected by the influence of this situational state on the *Working-Self goals*. Somehow, this method can be argued to be an approach to the *AGSys agency*: it is endowed with capabilities in order to act in a given way under self-referred circumstances.

5.8.6 Generalization

We accept that emotional capability requires ability for projecting, assessing and selecting courses of action. Consequently, as far as a system is able to perform complex reasoning, the emotions will result in more complex realizations depending on how the system is featured regarding intelligence (see rationale ♣ 54). From this perspective, emotion constitutes an essential feedback which inclines the system to measure and maintain its well-being equilibrium, regardless the level of complexity of each agent. That is, emotions provide the cognitive processes with quantifiable meaning and cognitive processes provide operation for wellness (see rationale ♣ 56).

Therefore, this reciprocal operation will be performed in a way as complex as the capabilities of the system allow. Highly complex systems (i.e. such as human beings) will deploy complex emotions, while simpler systems will deploy a range of more basic emotions related to their specific requirements. Therefore, the system *AGSys–ESys* should provide quantifiable meaning regardless the complexity of the agent.

By considering the work of *AGSys–ESys* as a *single machine*, this approach can be generalized for any type of system, whatever architecture it implements (see rationale ♣ 60). To give an example, it might be the case of a layered architecture which grows in complexity in a way to improve the ability for reasoning. These types of architecture are used to implement the physical layer at the bottom layer, and increase in complex reasoning—from scheme to deliberative—while moving towards the top layers. This thesis argues for single solutions instead of large solutions regarding system problem domains. Consequently, under the circumstances of a layered architecture such as the one herein described, the attention of each level is conceived an individual problem domain.

Under these circumstances, the use of the conceptual unit *AGSys–ESys* can be described by the repetition of this unit as many times as layers are deployed by the system. That is, an architectural design that displays a repeating transversal pattern between each layer, and the layer above it as illustrated in Fig. 5.53.

Results of an emotional assessment do not always correspond to the sensory evidence received during the observation. There are assessment dimensions that affect different levels of conceptualization and that, however, cause emotion from the source of a sensory evidence. According to Ellsworth and Scherer, there are appraisal dimensions that perform at the three levels of abstraction to which he refers (i.e. sensory, scheme and deliberative levels). The appraisal dimension of *novelty* is argued to be present within these three levels.

However, this is not a generalized rule that should affect all appraisal dimensions, and a number of them which only perform at specific levels of abstraction (such as *copying*) are argued. This generalization of a layered architecture together with a repeated *AGSys–ESys* single-machine, allows for defining specific emotional dimensions at each level of abstraction.

This results in the possibility of specifically designing emotional assessments related to every level of abstraction. Consequently, it gives rise to more complex artificial emotion that emerges from a distributed performance of computational emotion.

Finally, it should be remarked that, within the framework of a cognitive system with higher levels of abstract thought, social emotions are easily described according to this perspective. It allows for designing the emotional dimensions that regard the social interaction at those specific levels within which this social interaction is performed.

5.9 Conclusions

Adaptive systems use feedback as a key strategy to cope with uncertainty and change in their environments. The information obtained from the interaction with the external environment into the control subsystem can be used to change four different elements related to control: parameters associated to the control model, the control model itself, the functional organization of the agent and the functional realization of the agent.

There are many change alternatives and hence the complexity of the agent's space of potential configurations is daunting. The only viable alternative is to achieve a

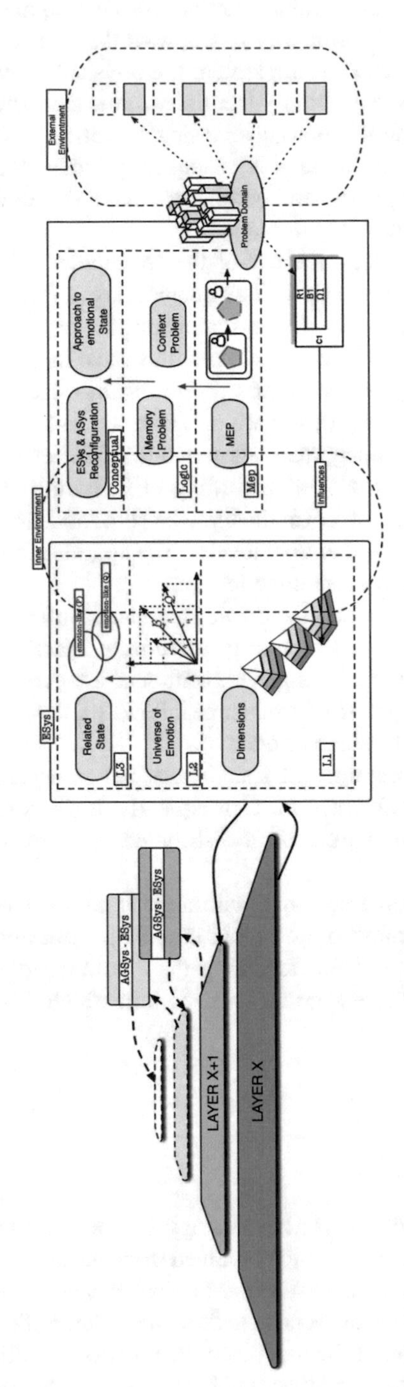

Fig. 5.53 General application of the strategy *AGSys–ESys* in a layered architecture

reduction of the dimensionality of this configuration space, and emotions play a critical role in this reduction. This reduction is conceived by means of an emotional model that monitors the inner state of the system with a predefined configuration for wellness. This allows the system to analyze its situational state and build its own representation regarding this state. Consequently, the system integrates emotion perception and the associated cognitive assessment opening new paths in order to achieve the challenge of autonomy.

A computational model intended to represent exploitable knowledge has been addressed, at every stage of the emotional assessment. There are a wide range of future works that emerge from this thesis, and many others that are required in order to achieve some parts of a future final model.

One essential question that needs to be analyzed before the end is the question of how *AGSys* identifies the object that is really causing the emotional effects that are being assessed. This is a complex question that needs to be deeply investigated and integrated in future developments. However, in this thesis the external object is considered as the unique object observed through the sensory system of *AGSys*, and consequently the unique recognized object by means of its ontological basis.

Besides this question, the one that refers to the perceptual process is also important. Even if several arguments regarding the perceptual operation have been introduced, this has not been addressed in its complete form in this approach. The recursive operation of perception integrates information related to the complete interaction of an external object within the system. Part of this knowledge is the emotional assessment, but this is not all. Under this circumstances, it is highlighted that perceptual recursiveness requires the study of the complete form that integrates not only emotion, but all the knowledge that emerges from the grounds of the inner circumstances of the system.

Additional questions will be deeply examined in the discussion presented in the following chapter.

Chapter 6
Discussion

The emotional phenomena explored in this thesis has been studied from a wide variety of fields of knowledge. These knowledge categories can be classified into two main groups: (a) the fields of psychology, cognitive science, neurology and biological sciences reviewed to obtain a better understanding of the emotional phenomena, and (b) several computer science branches such as Autonomic Computing (AC), Self-adaptive software, Self-X systems, Model-Integrated Computing (MIC) or the paradigm of models@runtime among others, in order to obtain knowledge about the various tools needed to design each part of the solution.

The final chosen approach derives from the entire acquired knowledge, and it is ascribed within the fields of Artificial Intelligence, Model-Based Systems (MBS), and additional mathematical formalizations to provide further insights as needed. This approach describes a reference model to feedback systems with valuable meaning, allowing for reasoning with regard to (a) the relationship between the environment and the relevance of the effects on the system, and (b) dynamical evaluations concerning the inner situational state of the system as a result of those effects.

With respect to the survivability of systems, this thesis strongly assumes that emotion will not provide all the necessary conditions to ensure a fully effective system. One of the proposals defended in this chapter will deny the statement of Wang [267]: The *fourth principle of systems science* asserts in its *7th theorem* that no system can achieve one hundred percent efficiency [267]. This principle affects the complete set of system capabilities, including survivability.

6.1 Introduction

This work is based on the search for correlation between the emotion in life systems and in artificial systems. The objective has not been to obtain a mirror model of the emotional behavior that life systems show, but the benefits that they obtain because

M.G. Sánchez-Escribano, *Engineering Computational Emotion—A Reference Model for Emotion in Artificial Systems*, Cognitive Systems Monographs 33,
DOI 10.1007/978-3-319-59430-9_6

of being able to behave in such a way. Life systems use feedback as a key strategy to cope with uncertainty and change in their environments. Once the need is detected, they change their functional strategy in order to face the situation and pursue the accomplishment of personal objectives.

Life agents are complex systems, and moving from one to another functional strategy could be an issue of great complexity without the work of emotion. Emotions seem to provide a reduction of dimensionality in the configuration space, providing in this way optimal strategies in order to select the required functional configuration at every moment. From this functional perspective, the objective of studying artificial emotion has not been a target of emotional expressions. It has rather been a question of the nature of emotional operation, and the search for principles that allow for optimizing cognitive processes by reducing the configuration space of systems.

This question has been comprehended in the following way: by assuming the final system and the emotional system as two different entities, the matter at hand has been: what are the foundations for their interdependence in order to provide analogous capabilities in artificial systems?

Life systems seem to use the emotional feedback in order to provide valuable meaning to relevances. Emotional feedback helps life systems understand the relationship between the environment and themselves, as well as the relevance of the impact of the environment on themselves. Even more, this valuable understanding seems to provide knowledge for decision making under circumstances of partial informational resources. Via this valuable information, emotions seem to enable life systems to a reduction of dimensionality concerning the environmental information. One essential objective of this thesis has been this reduction of dimensionality, and has been solved by controlling the inner states of the system instead of the external state of the environment.

Even more, this valuable meaning seems to provide knowledge for decision making under circumstances of partial informational resources. Emotions seem to provide the referred to reduction of dimensionality through this valuable meaning, and one essential objective in this thesis has been the construction of this quality and the means for its use.

This quality however is not the only exploitable knowledge that emotion provides. Through emotion, the system is able to describe the same reality with two different representations: the representation that refers to the logical reasoning, and the one that refers to the conclusion reached on the basis of the emotional assessment. This has been conceived by linking two representations in *AGSys*: the representation that refers to the ontological meaning (i.e. the *inner-Object*), and the one that refers to the emotional assessment (i.e. the associated *c-Images* that result from the original *emo-inner-Object*).

This chapter describes a summary about the approach proposed in this thesis, and the essential pillars that ground the solution. Afterwards, a discussion about applicability and the influence on autonomy and adaptiveness of systems will follow.

6.2 The Approach

The approach proposed in this thesis is not considered to be a model of the real phenomenon of emotion. The objective was not the theorization about the principles of the real phenomenon, but about the principles that support the analogy in artificial systems concerning the grounds of behavior. That is, it would be desirable that emotion theories might be explained within the framework of this proposal, and might fit well within the proposed framework.

According to Zhang et al. [276], there are six critical aspects by which emotion should be represented: (a) even when models are associated to a set of emotion types, they should be adaptable in order to explain additional emotions, (b) models should allow for simultaneous emotions, (c) models should allow for different emotions in two systems under the same context, (d) system needs the ability of mapping emotional situations into rules, (e) models should allow for "other systems emotional representation", or even more, of representing emotions of the group, and (f) emotional responses should emerge from the source of emotion and system goals.

Clearly, the solution proposed in this thesis describes the required correlation with these six principles:

1. *Explain additional emotions*
   The conceptualization of a *Universe of Emotion* allows for the representation of emotions within an algebraical space, by means of vectors denoted as *Emotion Attributes*. This way, a concrete emotion denoted as *happiness* can be described by a specific configuration of these vectors. And it can be differentiated from another emotion denoted as *sadness*. However, this universe enables additional states of the *Emotion Attributes* that are not denoted by a concrete label, but that are influencing the state of *AGSys*. The emotional assessment depends on the situational state of the emotional goals, and the work of *MEP* is triggered by this feature rather than by the recognition of a concrete emotional state.
2. *Simultaneous emotions*
   Simultaneous emotion is considered according to the assumption of systems with high levels of reasoning, cognition or complexity in such a degree as to allow for this simultaneity. Under this assumption, the entire operation of *AGSys–ESys* is conceived as a single-machine that operates on a part of the system. With the assumption of generalization (as illustrated in Fig. 5.53), the complex system is described within a layered architecture. This configuration assumes the work of this single-machine at every level of abstraction. This results in the definition of a dedicated *Universe of Emotion* at each layer, and the simultaneous work of the amount of universes within the whole system. Consequently, even in the case of two opposite emotions that cannot coexist within the same universe (i.e. simultaneous happiness and sadness), they however can be deployed at two different layers because they are being deployed at different *Universes of Emotion*. To give an example: it allows for sadness regarding the social environment, and happiness regarding the accomplishment of a personal mission.

3. *Two systems, same context and different emotions*
   The conceptualization made concerning the structure of the emotional goals allows for two essential features: (a) the structure of the emotional goals is conceived as an agent-based model, allowing for independent behavior related to the individual circumstances, (b) the integration of dynamics within this structure allows for a different evolution of the emotional goal under the identical circumstances of error in two different systems. Two systems might be detecting the same error under the same environmental circumstances, and even in identical situational states. Each system however, defines its own structure of emotional goals. And these goals are associated to a different dynamic of response. Consequently, the structure of emotional goals changes differently since each system's tendency is different. Additionally, even in cases in which the dynamics of goals would be identical for two systems, the structure of an agent-based model allows for independent behaviors related to the interaction (of individual units) within their frameworks.
4. *Mapping emotion into rules*
   The conceptualization of the *Working-self goals* as a reflection model as shared knowledge for an adaptive engine, allows for self-repairing the emotional goals in unstable situational states. According to the description provided in Sect. 5.7.2, Chap. 5 and illustrated in Fig. 5.36, actions for repairing are assumed according to cognitive and behavioral actions of *AGSys*. And it directly result in the mapping of rules concerning the emotional situation.
5. *Grouped emotional representation*
   This approach allows for the emotional individuality of each artificial system under identical conditions within a specific environment. This allows for the analysis of the emotional representation regarding each individual and regarding the group.
6. *Emotional responses on the basis of system goals*
   According to the description provided in Sect. 5.7.2, Chap. 5, and illustrated in Fig. 5.36, the conceptualization of the adaptive engine allows for behaviors related to the situational state of the *Emotion Attributes*. This way, emotional responses are obtained from the basis of the emotional goals and their specific situational state.

Regarding the emotional agent, emotions make sense only regarding cognitive competences of the agents (i.e. those minimums of rationale argued in Chap. 5). Our model-based cognitive *AGSys* is provided with a final pattern of data usable within the context of an emotional ontology. This provides *AGSys* not only with mechanisms in order to emotionally behave, but also to represent its own situational state in a high order level. *AGSys* is enabled to represent its own priorities at runtime, allowing for higher processes of reasoning. It is remarked that the work related to the *Working-self goals* results in effects on *behavior* and *cognition*. However, representation is a higher capability that requires to be solved at upper levels on the basis of the obtained patterns of data.

6.3 Emotion Based Autonomy

As argued in Sanz et al. [231], autonomous systems engineering is looking for functional control architectures that could augment the performance and resilience of technical systems operating in open-ended environments. And the inspiration from emotion is related to the search for mechanisms to achieve the levels of robust autonomy that animals demonstrate.

We might argue without risk of contradiction that, probably, the main objective of a biological being is to maintain its survival. Darwin [56] postulated that the way each biological being has evolved, either animal or plant, is related to the way in which their structures had to adapt to the environment where they live.

About autonomy In some sense, biology evolves in order to allow autonomous behaviors that endow biological beings with the best possible survival capabilities concerning the environment within which they move in. The notion of *autonomy* states a wide range of questions and problems that still do not have a complete answer. The most general idea of *autonomy* is the conception of freedom from external control or influence. This general idea of autonomy applies to plants as well as animals. Even more, it applies to any sort of artificial system that can maintain behaviors without external influence. According to this idea we could argue that we already have artificial autonomous systems.

From the point of view of engineering, *autonomy* is commonly referred to as the quality of systems to independently behave while pursuing mission objectives [106] (even though there are open issues still recognized). However, from our common perception, *autonomy* seems to refer to the capability of thinking and consequently acting for itself, rather than just being free from external control.

Our current artificial systems might be featured as ‘mentally handicapped’ systems. They still require external actions aimed to implement their autonomous capabilities. They are not capable of managing resources on their own, in order to change their rules of behavior under different circumstances from those considered by design.

A glance into nature provides a different understanding of *autonomy* depending on whether it concerns plants or animals. Even when both of them have evolved according to analogous goals of survival (see Darwin [56]), their autonomous behavior seems to be quite different. Plants can react to the resources, threats and traits of their environment, but they do not have the capability of changing their environment when it is not adequate for their needs. However, animals can do it. Animals can react to their environment and not only react concerning resources, threats and traits, but also to move in and to move throughout their environment and search for optimal conditions concerning their needs. Maybe this is the reason why animals have evolved to approach this different type of autonomy. As they are allowed to move from place to place, they have evolved to analyze the environment concerning needs, realize what they require to solve each issue and search for what is required to solve it.

Even when this hypothesis cannot be a demonstrable theory without additional arguments, it offers an outlook of the principles that govern *freedom autonomy*. Freedom in autonomy is related to the way animals realize the information that comes from their environment in order to identify what they have available to solve the challenge. In case they detect lack of resources for working, they move the system (i.e. move themselves) towards those new places in the environment within which they can find what they need. Note that this hypothesis integrates social behaviors by means of which agents might interact with other agents in order to fulfill required resources.

In the end, we engineers might conceive survival as *the state or fact of a system of continuing its systemic equilibrium, concerning circumstances that threaten its integrity*. This target of equilibrium is a different target from those related to system mission.

Thesis 18 ♣ *Autonomous System*
An autonomous system is the system that works with the meta-objective of continuing its systemic equilibrium.

Emotion and autonomy Downey describes how complex systems of agents show properties as a whole that their components do not. He describes some examples such as the one of the agents in the Schelling's model: they are not defined as 'racist', but their interactions result in behaviors that look as if they were [65]. The suggestion is that it might be an approach to questions related to consciousness and free will. From his point of view "Free will is the ability to make choices, but if our bodies and brains are governed by deterministic physical laws, our actions would be determined" [65]. Somehow, this perspective assumes that life beings are distributed complex systems formed by units that can be conceived as agent-based parts.

After showing the arguments of William James and David Hume regarding this question of free will, he accepts the agreement concerning the impossibility of having free will if the parts are deterministic. The main suggestion made by Downey is the alternative that "free will, at the level of options and decisions, is compatible with determinism at the level of neurons (or some lower level)". And this argument inspired the study according to the vision of a lower level unit that might allow for *Emotion Based Autonomy*. And this lower level unit has been conceived in the form of units (i.e. *c-emoGoals*, *c-Concerns*, and *a-Dimensions*), that all together define the structure of the emotional goals.

This perspective however requires additional considerations. According to the viewpoint of this thesis, one essential aspect missing in control is the lack of *Working-Self Goals* as a critical feature of independence. This thesis argues that one key feature of *autonomy*—which consequently will affect the control for *autonomy*, is the structure and management of the *Working-Self Goals* structure. We hold that control of discrepancies within this structure, is a key corner in *control for autonomy*. The operation of managing *Working-Self Goals* comes to solving discrepancies when external events appear, and these external events can be related to either environmental changes, as well as to commands to complete tasks.

Regard environmental changes or commands to complete tasks.

This way, emotion and cognition are integrated within the same architecture, improving this way the competences of the agent to (a) control its inner goals (i.e. personal goals or emotional goals), and (b) perform real-time adaptation mechanisms.

Thesis 19 ♣ *Emotion and Autonomous Systems*
Emotion is the underlying principle for autonomous systems.

6.4 Adaptive Systems

This section will analyze the context of adaptive systems, and the influence of emotion on reconfiguration and adaptiveness capabilities. Somehow, emotion can be conceived as an engine that generates meta-directives towards the adaptiveness. However, the idea of emotion-based adaptiveness should be clarified.

Questions about emotion-based adaptiveness Undoubtedly, a lot of adaptive responses can be noted in biological systems and some of them are related to reactive rather than to the emotional functionality. To give an example, the sensorial perception of excessive warm in your hands (i.e. a sensation of being burnt), immediately results in the action of removing your hands from the source of heat. An adaptive response can be considered rather than an emotional behavior. Even if the emotional response related to this event is concurrently triggered, the adaptation to the environment is obtained by means of a reflexive response. The performance of emotion under this circumstance is related to processes of learning for future actions.

If emotions were to be the only relevant explicatory variable of system adaptation, we could then conclude it has malfunctioned. The reason is that adaptation does not work always and we en up dying. The truth is that many factors help represent reality and these affect the capability to adapt. Emotions happens to be just one of these factors.

Self-adaptive systems Di Marzo Serugendo et al. [63] present a relevant roadmap in adaptive-systems research. This work conceives self-adaptive systems as those which are able to adjust their behavior in response to their perception about the environment and themselves. This feature is comprehended according to several aspects such as user needs, environmental characteristics, etc. And the design is related to the analysis of four dimensions (a) the dimensions associated with self-adaptability aspects of the system goals, (b) those associated with causes of self-adaptation, (c) the mechanisms associated to achieve self-adaptiveness, and (d) the effects of self-adaptiveness upon a system [63].

Goals are objectives that the system should achieve, and that pose questions about the following dimensions: evolution within the lifetime of the system, flexibility in the way goals are expressed, duration or validity of a goal throughout its lifetime, multiplicity or the number of goals associated with the self-adaptiveness functionality, and the relation of dependency among the objectives together with the subsequent questions that this last question presents [63].

Changes are conceived as the party which causes the following requirement of adaptation, and the source of these changes is accepted to be either external or internal to the system. Context is conceived as information that is computationally accessible and available to be exploited. Related to changes are several dimensions such as the source which is the origin of the change, the nature of change, the frequency in the case of repetitions, or the capability of the system to predict change [63].

The mechanisms—by means of which the reconfiguration is achieved, present questions about the type of reconfiguration (parametric, functional, structural, etc.), the autonomy of the system in order to attend the optimal reconfiguration, the organization and the scope of the adaptation, and questions about the time to reconfigure and trigger forms [63].

Finally, it is essential to take into account the subsequent effects related to an operation of self-adaptiveness. It is important for the success of the system to devise the criticality in case of fail in the adaptive processes, predictability of these effects in value and time, to envisage possible negative effects, and the persistence of service delivery when facing changes (i.e. provide resilience and justify it) [63].

Emotion and the self-adaptive systems design This thesis focuses on achieving analogous competences in artificial systems as those provided by emotion in life agents. As it has been argued, emotion is not conceived as the essential feature for adaptiveness but as an essential feature to positively influence the process of adaptiveness. What is a key feature of emotional systems is that they are scheduled to give priority to the self-goals. And it results in processes that are intended to protect the agent, to ensure its survivability, to achieve optimal interaction with the environment, etc.

That is, the work of emotion is focused on the modulation of the system in order to behave in such a way as to ensure the accomplishment of its self-objectives. With this conception, a model of emotion will influence the dimensions of self-adaptiveness for this target of self-accomplishment:

1. *Goals*
   The structure of emotional goals proposed in this thesis is controlled by an adaptive engine that ensures the accomplishment of *Working-self goals*. And this is how *AGSys* understands its own objectives. Additionally, the proposed agent-based model defines every emotional dimension as influenced by dynamics that come from multiple sources: (a) the dynamics of every unit, and (b) the dynamics from the units interaction.
   One of the challenges regarding goals is the requirement of performing a trade-off analysis between several potentially conflicting goals [63], and emotion seems to be a promising option in order to use system preferences to compare situations under *"Pareto optimal conditions"* (by paraphrasing authors in Di Marzo Serugendo et al.).
2. *Change*
   The monitoring of *AGSys* is directly related to the quality of service within the inner environment of the system, concerning the requirements of emotional goals. The interaction of these emotional goals is controlled by the *Working-self goals*,

therefore the requirement of accomplishment of these emotional goals might be also controlled. Since the accomplishment of emotional goals can be comprehended as qualities of service required by the system itself, emotion might become a promising tool in order to monitor the system according to several QoS properties of interest.

3. *Mechanisms*
   The distribution of goals described in this thesis allows for action readiness and basic regulation loops that can act at every moment, whenever they are required. The whole emotional assessment continues, but the essential needs of fast reconfiguration are accomplished at each moment they are detected. The final assessment is also used in order to reconfigure at different levels within *AGSys*. This is a distributed mechanism that improves the design of self-adaptive systems: according to Di Marzo Serugendo et al. the application of *"the centralized control loop pattern to a large-scale software system may suffer from scalability problems"*.
4. *Effects*
   One key characteristic of life systems, is the use of emotion to assess a specific scenario integrating the past (i.e. emotional-based memories) and the future as it applies to current actions. It is like a *Meta PID control* that acts controlling a meta plant of effects within *AGSys*. Di Marzo Serugendo et al. speak of the need to develop more advanced and predictive models of adaptation, in case of failure in the success of requirements due to the side-effects of change. Therefore, emotion might become a good option in order to overcome such a challenge.

Emotion and requirements in self-adaptive systems This is concerned with the question of flexibility in evolving system-requirements at runtime under relevant circumstances related to unknown environments. It is not possible to monitor the entire environment of a system so, consequently, environments always integrate uncertainty. A matter at concern is which aspects of these environments might become relevant for adaptation at runtime, and how to monitor changes related to these aspects.

This thesis proposes a solution that monitors the inner environment of the system. It allows for monitoring the wellbeing of the system, rather than monitoring the possible effects on the wellbeing of the system. This involves an improvement concerning the attention to uncertainty. The system does not need to recognize the environment a priori: when inner failures are detected, the system deals with the problem at runtime. Immediately, the system will improve its capabilities to recognize the current environment and rebuilding the scene. Emotions state a different paradigm concerning the environmental uncertainty and the self-adaptiveness. It is not the comprehension of the environment that is required in order to trigger actions, but the comprehension of the inner state.

This new paradigm however, does not imply that the system will always recognize which action or actions should be triggered at runtime. In the case of lack of responses, the system might acquire the state of self-defense. This is a different state from self-adaptiveness, but most likely will respond with associated actions that might improve the resilience of systems.

This solution allows for requirements reflection. This is a key question argued by Di Marzo Serugendo et al. regarding the improvement of self-adaptiveness. From the perspective of authors, reflection *"enables the system to observe its own structure and behavior"*.

The last stage in the *MEP* process herein proposed, enables *AGSys* to observe its own structure of *Working-self goals*. According to this approach, this structure is the one which ensures a correct state of the emotional *a-Dimensions* (on the basis of monitoring the *Emotional Attributes*). These emotional attributes—in the form of vectors, are transversally capturing different aspects of the created environments for every emotional dimension. Hence, relevant aspects of system wellness can be monitored and analyzed under the constraints imposed by the own system wellness (its own reflection model related to wellness).

Emotion and assurance in self-adaptive systems The roadmap on the self-adaptive systems researched by Di Marzo Serugendo et al., describes the feature of systems assurance related to the verification of activities when systems adapt. From their perspective, adaptive-systems are highly dependable on context changes and the subsequent adaptiveness decisions. And they require verification activities to provide continual assessment. They then require moving the *Verification and Validation strategies* from the design and requirements stages, to the operational stage using runtime assurance techniques.

The argument by Di Marzo Serugendo et al. is that 'properties within a system' (i.e. goals), can vary in their importance from context to context. And this results in additional derived requirements concerning design: (a) dynamic identification of the changing [system] requirements, (b) adaptation-specific Model-driven environments and (c) agile runtime assurance (the last question of *"liability and social aspects"* argued by the authors will not be addressed here because it does not apply to the objectives of this thesis).

1. *Dynamic Identification of Changing Requirements*
   It concerns the change of requirements as a result of a change in the context. This perspective however, searches for context identification and the need of utility functions in order to analyze trade-offs between the goals aligned with the context. And they propose probabilistic techniques in order to avoid the problems caused by the intrinsic uncertainty.
   However, probability depends on the extent to which something is probable, and on the previous definitions of those likelihoods as related to the defined cases. The emotional approach herein proposed allows for a different paradigm as explained earlier. The system does not need to recognize the external environment a priori. By means of the conceptualized *inner-Environment*, when inner failures are detected the system will deal with the problem at runtime. Immediately, the system will improve its capabilities to recognize the current environment and build the scene. And it is then enabled to build its own likelihoods for future similar events.

2. *Adaptation-Specific Model-Driven Environments*
   It is related to the capability of Model-based systems to exploit models and provide estimations of systems' status in order to predict the possible impact of a concrete context. The solution argued by the authors is the use of descriptions of uncertainty attributes by using probability distributions, attribute value ranges, or historical values of these attributes among their proposed examples. The emotional approach allows for runtime building of these prediction models, at the same time that they provide self-evaluation of the situational state. Consequently, this approach provides the required means in order to keep runtime models synchronized with the changing system (a key requirement hold by the authors).
3. *Agile Runtime Assurance*
   It regards the requirement of runtime verification within efficient agile solutions which deal with space/time complexity. Di Marzo Serugendo et al. argue that, when formal property proofs (i.e. goals proofs) *"do not seem feasible, runtime assurance techniques may rely on demonstrable properties of adaptation, like convergence and stability"*. Adaptation refers to a transient behavior, and the desirable attainment of final stability concerning this transition of states. The emotional approach provides a space of emotional goals stability within which mission goals can be controlled focusing on a global target of system wellness. However, the authors highlight the difference between 'desirable' and 'correctness' concerning the uncertainty of designers about what is a correct adaptation in unforeseen contexts. This is a key question that emotions might help resolve: whether it is the system, rather than the designer, that which decided about its own correctness. The designer's role is then to decide on 'desirability'. We argue that this might be achieved through the conceptualization of emotion.

The overall conclusion regarding the influence of emotion in self-adaptation is that, as outlined by emotion science, it is an essential cornerstone for adaptiveness. As justified in this section, emotion has influence (a) on modeling dimensions in order to obtain precise models for runtime reasoning, (b) on techniques in order to deal with uncertainty, and (c) on the improvement of systems assurance.

Thesis 20 ♣ *Emotion and Self-adaptive Systems*
Emotion is an essential feature upon which self-adaptiveness depends.

6.5 What Is Artificial Emotion

Life systems are provided with a sensorial system that enables them to perceive and measure the environment within which they live. Commonly, five senses are accepted through which they perceive a whole range of physical changes and take information about the world related to physical things. These senses provide systems with information which is used in their interaction with the environment. However, explanations about emotion are not as easy when we refer to the informational processes the brain involves. The reason is that the whole perception mechanism includes not only the physical signals that come from the senses, but also subsequent analysis that emerge from the basis of essential cognitive processes. But the brain lacks analogous sensors that measure informational processes in a direct way. Hence, there is no way to directly detect the inner informational processes into the brain.

Emotional goals are meta objectives aimed to ensure optimal systemic equilibrium. Thus, emotional response is only a part of the whole functionality of emotion (i.e. a subsequent reaction to specific states related to the value of the emotional goals). The operation of emotion additionally shows an image of this equilibrium to the system in order to enable the system to represent its own state and exploit its resources to optimize it. Therefore, the operation of emotion determines system behavior resulting from changes related to the systemic equilibrium (either in positive or negative ranges). And the systemic equilibrium is controlled through the accomplishment of the emotional goals.

Actions are not triggered without previous causes. Physical cues coming from the environment are measured through the senses, and this enables the system to deal with the external environment. Following an analogous reasoning, emotional cues coming from the inner environment need to be measured to enable the system to deal with its own state. Both cases are featured by the triggering of actions, when incongruences in the relationship between the system and the environment are detected.

Emotion is made up of a set of connected functional particles forming a complex whole, which aims to produce movement inside of living beings when required. The source of this movement (i.e. the cause that triggers actions) is the differential range between what is expected and what actually happens, with respect to the balance in the state of the system (regardless any range's characterization).

According to the approach proposed in this thesis, the emotion appraisal provides energy-like quantities. Under the conceptualization of this thesis, these quantities are related to the amount of energy per objective at a specified appraisal dimension. This solution, results in the emotional pipelines that supply information about the loss and gain of value in emotional objectives. This differential between the expected value of an emotional objective and the value it actually has, is hereby assumed to be the source of movement to attend system requirements concerning this emotional objective.

The systemic equilibrium is influenced by the operation at several levels of abstraction. If the referred to systemic equilibrium is denoted as system wellness, emotion seems to be like a power-unit aimed to trigger meta-orders in order to stabilize patterns of wellness within each appraisal dimension. That is, a meta-adaptive engine intended to endow transversal-adaptiveness through several levels of abstract operations within the system. Additionally, since we accept this transversal operation of emotion, the operation of this meta-adaptive engine is assumed as being mainly distributed.

The prefix *'meta-'* features the adaptive engine in such a way so that it constrains the features of the system within which it is hosted. This *'meta-'* denotes something of a second-order or higher kind. Consequently, the physical layer of a system cannot implement this meta-adaptive engine of emotion since there is no layer below. From this perspective, it is argued that emotion can only appear in systems that implement some type of rationale (as argued in Chaps. 2 and 5).

From this perspective, the final objective of emotion is the transformation of some type of information (i.e. coming from the environment) into another type of information that is relevant to the system's wellness so it can measure this wellness and react to it.

Thesis 21 ♣ *What is Artificial Emotion*
Emotion is a meta-adaptive distributed engine intended to cause transversal-adaptiveness, in order to provide the means to control the systemic equilibrium of the systems.

Generally speaking, emotion in artificial systems is related to the ability of a system to determine its state and its behavior concerning its systemic equilibrium. The systemic equilibrium defines a number of variables that are regulated by the emotional goals. And the emotional goals supervise the evolution of these variables.

From our perspective, emotion appears because of the need for feedback in biological systems. When life systems evolve in their capability of cognitive reasoning, the interaction with the environment becomes more complex than before. Therefore, the sensorial feedback alone is not enough to provide life systems with informational resources in accordance to their evolved capabilities.

Finally, we argue emotion to be (a) a foundation for autonomy, and (b) a cornerstone for adaptiveness.

6.6 Conclusion

As argued in previous chapters, from an engineering perspective there are differences between the emotion-recognition research, the emotion-expression research and the on-feeling research. This engineering perspective aims to be the framework within which the three threads might be designed. Artificial systems can be designed according to emotional descriptions that, to give an example, might not be detected by humans. This is one of the key dangers that future works on emotional systems might represent for the human being.

However, the solution herein this thesis described allows for modeling emotions by means of a functional analogy to the biological operation. And it might improve the system's design strategies. What is defined at design time, are a series of principles analogous to those of the reptilian brain. The objective of this choice of design is to enable the system to understand its own requirements for survivability. This way the system: (a) can represent its situational state, (b) is enabled to learn what is good or what is bad from its own experience, and (c) can express its own artificial-emotion according to its own experience and learning processes.

Chapter 7
Future Works

The aim of this thesis is to investigate emotion in life systems, from the perspective of the artificial systems. The main object of analysis has been to determine the correlation of functionalities, that is, to ascertain the particular use for which artificial emotion might be designed. Therefore, the key goal was not to produce a mirror model based on the emotional behavior that life systems show; But rather, to identify the benefits that life systems obtain because they behave in such an emotional form.

The final approach has been described within the fields of Artificial Intelligence, Model–Based Systems (MBS). Also, additional mathematical formalizations have been provided to ensure the of specific issues as needed. Much of the future work to be developed in this subject is related to the limitations of this thesis and, particularly, to aspects which have not been covered here.

Furthermore, even though our approach has been sufficiently justified on the basis of the existing literature, verification of the proposal requires direct testing. There are some constructs that are essentially delineated and that entail additional research. Examples include the topological form of the recursive function of perception, the final implementation of *R–Loop* and *MEP*, or putting into effect an emotional ontology to convey emotional value.

Besides these, the complexity of the emotional phenomenon imposes additional considerations about the results herein proposed. These considerations do not only refer to the question of the general applicability of artificial emotion on the basis of this approach, but to additional questions about the explanatory capability of this approach in order to verify emotional theories.

Consequently, this thesis assumes three broad lines of future works: (a) those related to the approached solution that still require to solve essential questions herein just delineated, and (b) future works related to the general applicability of emotion to artificial systems on the basis of this theory, and (c) the explanatory potential of this approach in order to verify emotional theories.

M.G. Sánchez-Escribano, *Engineering Computational Emotion—A Reference Model for Emotion in Artificial Systems*, Cognitive Systems Monographs 33,
DOI 10.1007/978-3-319-59430-9_7

7.1 The Approach

This approach describes a reference model to feedback systems with valuable meaning, allowing for reasoning with regard to (a) the relationship between the environment and the relevance of the effects on the system, and (b) dynamical evaluations concerning the inner situational state of the system as a result of those effects. It has been conceived under a multi-purpose architecture that implements two broad modules in order to attend to: (a) the range of processes related to the environment affectation, and (b) the processes related to the emotion perception-like and the higher levels of reasoning.

This thesis has applied an analytical method to study, from an engineering point of view, the substance of common elements within the various emotional theories. Albeit faced with contradictory finding or incomplete descriptions, our focus has been on those aspects of the general theories that provide a general explanation of the emotional function. Our goal was to identify an engineering correlation in the common perspectives of those theories in order to assess their value in our context of engineering science. These correlations have been used to explain every part of our proposal.

This approach is conceived under the general principle of describing a framework that requires an emotional theory within which to obtain a final ad-hoc implementation. That is, we do not focus on a specific single theory, but rather on any theory that might be chosen.

Continuation of the Approach

Following from our previous section, we then consider the next steps needed to implement this approach. Here, the essential objective is to verify the model and to this end, the framework requires an initial implementation to obtain a first preliminary version. This version will provide information with respect to the failures and lackings of the reference model and its foundations, as well as about some results which have not been predicted earlier in this work.

Clearly, it is derived from our analysis that an emotional theory is required to implement this framework. Here, a method of emotion construction has been proposed. However, the emotional bases under which the emotional objectives need to be built require the analysis of emotion concerning the final system. And this requires for an emotional explanation under a scientific methodology.

The definition of every *a-Dimension* and its related *c-Concerns* and *c-emoGoals*, requires the selection of an emotional theory that explains the emotional principles under which an agent describes its states of affect. And this emotional theory (that will be supported by a scientific method), will help in the construction of the *inner-Environment*.

Limitations of the Approach

This thesis proposes a possible type of the perceptual function. This function might be assumed in such a way that it maintains its form while concurrently integrates new abstractions. Another proposal is the possibility of defining an invariable topological type that might integrate at the same time the growing of the data, even under possible forms of fractals (i.e. continuous changes that can be continuously reversed).

To formalize this, additional studies, analysis and research are deemed necessary. This conceptual line is argued as an important issue to study in order to obtain responses about an artificial function of perception. Once formalized, the following challenge is its computational implementation.

Required Artifacts

In order to manage cognition and behavior in the emotional state, a reflection model of *Working-self goals* under the perspective of Pleydell-Pearce [50] is required. This provides shared knowledge to adapt those requirements related to the emotional dimensions to those other requirements related to the systemic equilibrium (i.e. systemic stability herein assumed as wellness). Consequently, it enables the curse of actions aimed to influence cognitive? based reasoning and behavior.

An additional requirement is that we build an emotional ontology in order to assess the entire data that comes from *MEP*. This way, *AGSys* might be provided with different dimensions of emotional meaning. This ontology needs to be constructed on the basis of patterns described by the data provided by *MEP*.

7.2 Emotion Research

> An agent is something that acts ('agent' comes from the Latin 'agere', to do). Of course, all computer programs do something, but computer agents are expected to do more: operate autonomously, perceive their environment, persist over a prolonged time period, adapt to change, and create and pursue goals. A 'rational agent' is one that acts so as to achieve the best outcome or, when there is uncertainty, the best expected outcome.
>
> Russell and Norvig [223]

Emotion has been argued as an essential feature for adaptiveness and survivability. However, it might be rather assumed as an essential feature aimed to transform the environmental information to ensure the system understands the environment on the basis of its systemic stability (i.e. system wellness). Thus, emotion provides the system with valuable meaning which might be assumed as implicit information.

Questions about 'Qualia'

> It can also be maintained that it is best to provide the machine with the best sense organs that money can buy, and then teach it to understand and speak English. That process could follow the normal teaching of a child. Things would be pointed out and named, etc. Again, I do not know what the right answer is, but I think both approaches should be tried.
>
> Turing [260]

Under an analogous point of view to this of Turing, it is best to provide the machine with the best structure to perceive emotion, and then teach it to understand emotion.

Under this perspective, artificial emotion should be designed to enable systems for creating their own emotional range. Since each system has its own references to asses the same environment, it establishes the principles for *Qualia*. The framework proposed in this thesis might serve as an starting point for future works related to these questions.

Uncertainty and Loops

The environment of the system and how it is modeled is always a critical factor for designers. An ideal situation might allow a completely defined environment at design time. However, this becomes an unachievable task of systems identification since the final model always depends on observations. Even more, the very same definition of *model* inherently implies deviations from real world (see Booch et al. [26]). So uncertainty, unstructured environments, and possible unpredictable events cannot be fully considered in the initial phases of a system design. Descriptive models about the reality are not defining the complete set of arbitrary changes of the environment and, consequently, prescriptive models depend on these models that not reflect correctly the nature of change. Nonetheless, the way in which unpredictable events are attended influences the final resilience and robustness of the system.

Since the uncertainty coming from the environment cannot be entirely managed, the system might allow the progress of the after-effects within the system towards a better point from where to monitor them. This is a new perspective of design for possible future works.

As designers, we can have a more complete insight about the operation and the internal structure of the system than about these issues as they concern the external environment. Thus, it might be appropriate to study new methods to model inner environments related to the accomplishment of essential requirements (i.e. relevant information for the system that the system will be enabled to use). Once established a new domain of operation, the system has to be endowed with the necessary causality to use the information that comes from it.

However it is essential to clarify a key aspect of the improvement of systems' adaptivity: perception is not a perfect provider of information when attempting to model the environment. We assume errors in our responses to the environment, and these errors might come from different sources: for instance, models of the environment erroneously built, exploited and/or interpreted among other. Even more so, we assume differences from our perception to the perception of other people regarding the same reality. Thus, since we are not perfect systems in our responses or in the way we model our reality, we cannot expect that our artificial systems might be ideal at all.

Autonomy

As it has been argued in this thesis, emotion is an essential cornerstone for autonomous behaviors. The essential capability of an autonomous system is its ability to change rules on the basis of its own inferences. However, this changeability of rules is not as easy as it might seem.

It is not a question of reconfiguring rules, but rather a question of creating new rules. This is an interesting field that might provide new insights about artificial intentionality in systems: the goal of providing purposiveness to systems.

This ability is not only based on the requirement of a causal model that relates to (or that connects the) the environment and the system. It is also a question of the effects that the environment produces, the evaluation of these effects, and the representation of the environment, effects and the evaluation of the effects by the system. This way, it becomes a exploitable model built on the basis of implicit information.

Glossary

A

Affect *(Thesis)* • Regarding an emotional system, it is the valuation that a system makes concerning some element in its environment (this element is denoted as object by [54]).

Action Readiness *(Thesis)* • State to which a system moves, in order to adapt its behavior to a critical environment and the subsequent requirements.

Action Tendenct *(Thesis)* • Propensity of a system to behave under specific forms, and that is intrinsically related to its intrinsic characterization.

Artificial Consciousness *(Thesis)* • It will allude to any type of information abled to be exploited by any rational system.

Artificial Emotion *(Thesis)* • (1) Related to the experience: general name used to refer the complete experience of emotion in some artificial agent. (2) Related to the broad meaning of the concept: Emotion is a meta–adaptive distributed engine intended to endow transversal–adaptiveness, in order to provide the means to control the systemic equilibrium of the systems.

Agent, Rational [223] • Fundamental to the approach to artificial intelligence. It is referred as an artificial agent that always maintains an interaction with the environment by means of sensors and actuators. Rational agents accomplish some type of performance measure in order to: (a) interact with their environment, and (b) assess consequences that concern each change caused on this environment. With this assumption, a *rational agent* is defined by accepting that *rationality* depends on four things: (a) the performance measurement, (b) the agent's prior knowledge, (c) the actions knowledge, and (d) the sequence of perceptions till the current time.

Agent, Error–Reflex *(Thesis)* • c.f. Error–Reflex agent.

Appraisal *(Thesis)* • Function aimed to estimate fitness and aptitude in key events for the system (a.k.a. cognitive–based emotional assessment). Under [5] it refers to a progression of events related by the work of perception that are responsible of emotion arising [5] (this term was introduced by this author).

M.G. Sánchez-Escribano, *Engineering Computational Emotion—A Reference Model for Emotion in Artificial Systems*, Cognitive Systems Monographs 33,
DOI 10.1007/978-3-319-59430-9

Appraisal Dimension *(Thesis)* • Each of the N dimensions of emotional assessment in a system. They are conceived under an analogous view to that of a vector space, and it is denoted by *a–Dimension*.

Architecture *(Emotion Science)* • Termino relativo a.

Architecture, Reference [257] • Set of principal design decisions that are simultaneously applicable to multiple related systems, typically within an application domain, with explicitly defined points of variation.

Architecture, Software [257] • Set of principal design decisions made about a system.

Appraised Subsystem *(Thesis)* • It is the system formed by a set of independent components inside of a system, whose interaction might become critical under the requirements of some emotional dimension (c.f. Definition 5.5 in Chap. 5).

Arousal [54] • It denotes the presence of signs of autonomic nervous system activation (...) which are reasonably covered by lay terms such as excitement.

Arousal, Computational *(Thesis)* • *Arousal* is the state of exciting by means of which some essential processes are triggered in order to optimally face a punctual situation, and constitutes the ground for emotion [5, 54, 107, 118]. Arousal state is conceived in artificial systems as interruption states, that is, a state in which alerts aware the system about the requirement of immediate attention regarding some source that is establishing high-priority.

Autonomous System *(Thesis)* • System that works under the meta–objective of continuing its systemic equilibrium. System featured by its ability to maintain its systemic equilibrium.

Autonomy *(Thesis)* • Ability of a system to generate its own rules and self–manage without external actions. When *autonomy* refers to agents, this generation of rules is grounded by the influence of its systemic equilibrium.

Aware [259] • The scanned symbol is the only one of which the machine is, so to speak, directly aware (c.f. Artificial Consciousness).

Awareness, System *(Thesis)* • Function by means of which a system recognize its own capability related to its state.

B

Behavior [259] • The operation of any machine, which is intrinsically determined by its configuration (i.e. arrangement in which artifacts are interconnected).

C

Chunked–Memories *(Thesis)* • Computational artifact that we situate inside or further in the system, and that represents solid pieces of memories as a source of knowledge for the system.

Cognition, Computational *(Thesis)* • Process by means of which the system acquires, update or exploit knowledge related to the relationship between the environment and itself.

Cognitive Architectuse [106] • Blueprint for intelligent agents. It proposes (artificial) computational processes that act like certain cognitive systems, most often, like a person, or that act intelligent under some definition. Cognitive architectures form a subset of general agent architectures.

Computational–Concern *(Thesis)* • It refers a collection of functional artifacts used to build a metric with which to measure a concrete *a–Dimension*. This measure is conceived in terms of positive or negative distance to those matters of importance for that *a–Dimension*, and is denoted by *c–Concern*.

Computational Emotion *(Thesis)* • Computational work aimed to obtain the complete set of artifacts required to build the emotional experience into some artificial agent (i.e. this experience is denoted in this thesis by Artificial Emotion).

Computational emo–like Goal *(Thesis)* • The value that uses the system to measure the positive or negative error regarding some *c–Concern* (it is denoted by *c–emoGoal*).

Computational–Image *(Thesis)* • It will allude to patterns of states concerning a set of system artifacts (it is denoted by c–Image).

Computational–Map *(Thesis)* • It will allude to patterns of relationships concerning a set of system artifacts (it is denoted by c–Map).

Computational Process of Emotion *(Thesis)* • Process by means of which the emotion–Object is built (denoted cp–Emotion).

Computational Process of Feeling *(Thesis)* • Process by means of which the emotion–Feeling is built (denoted cp–Feeling).

Consciousness, Artificial *(Thesis)* • c.f. Artificial Consciousness.

Context [62] • Any information that can be used to characterize the situation of a software entity.

Context, Bounded [74] • BOUNDED CONTEXT refers the delimited applicability of a particular model.

Control *(Thesis)* • Process during which are taken actions to achieve specific effects on a concrete system.

D

Domain *Computer Science* • A sphere of knowledge, influence or activity (c.f. glossary in [74]). Also the problem space that defines characteristics, vocabulary, motivation (why this domain exists), etc. (c.f. [257]).

E

Embodiment *(Thesis)* • Attribute of a system that imply physical realization (it may refer to either realization of processes or body structure).

Emotion, Artificial *(Thesis)* • c.f. Artificial Emotion.

Emotion, Computational *(Thesis)* • c.f. Computational Emotion.

Emo–Inner–Object *(Thesis)* • Computational artifact situated inside or further in the system, which represents those relevant changes that an external object causes on the system concerning emotion–based references.

Emotion–Object *(Thesis)* • Computational artifact that we situate inside or further in the system, which accumulates the information about the emotion–based consequences that an external–Object is causing on the system.

Error–Reflex agent *(Thesis)* • Software agent aimed to monitor the inner state of *AGSys*, concerning the requirements imposed by some emotional dimension.

Environment *(Thesis)* • General term used to refer the surroundings or conditions in which a system operates.

F

Feeling–Object *(Thesis)* • Computational artifact situated inside or further in the system, which represents an evolution of the *emotion–Object*.

G

Goal *(Thesis)* • Feature liable to vary that constraints the transition of systems from a current state, to another state related to the accomplishment of a required value of that feature.

I

Information *(Thesis)* • It refers in artificial systems to exploitable knowledge.

Inner–Environment *(Thesis)* • Software–based description in order to obtain an environment within which *ESys* may accomplish its missions, related to emotional requirements. Consequently, it will be constituted by those values that *Error–Reflex agents* provide.

Inner–Object *(Thesis)* • Computational artifact situated inside or further in the system, which represents descriptive aspects of the external objects as a source of knowledge for the system.

K

Knowledge *(Thesis)* • General term used to denote the set of exploitable models that a system integrates.

M

Model *(Computer Science)* • Model is a representation of a real world process, device or concept (c.f. IEEE Standard 1233–1998 (R2002)]). Also, model is an approximation, representation, or idealization of selected aspects of the structure, behavior, operation, or other characteristics of a real-world process, concept, or system (c.f. IEEE 610.12-1990).

P

Perception *(Thesis)* • Generally speaking, it refers a recursive process of information integration. Specifically, this information is related to the sensory acquisition plus the effects caused within the system and the relationships among these effects.

R

Recursive Loop *(Thesis)* • Recursive functional artifact, aimed to provide an explicit interpretation of the object by means of successive executions of the same conceptual loop (it is denoted by *R–Loop*).

S

Scenario *(Computer Science - IEEE Std 1362-1998)* • (A) A step–by–step description of a series of events that may occur concurrently or sequentially. (B) An account or synopsis of a projected course of events or actions.

Somatosensory System, Computational *(Thesis)* • Set of connected interruptions triggered by a system.

Survivability, Artificial system *(Thesis)* • The state or fact of a system of continuing its systemic equilibrium, concerning circumstances that threaten its integrity.

System *(Computer Science - IEEE Std 1362-1998)* • A collection of interacting components organized to accomplish a specific function or set of functions within a specific environment.

T

Tendency, Emotion *(Thesis)* • Inclination of a system to respond under specific emotional characteristics. In artificial systems this is defined by design and potentially modified by means of machine–learning processes.

V

Valuable–State *(Thesis)* • Denoted by (v–State), it refers to the computational artifact which represents a midway step that acts as an intermediate stage between the *emotion–Object* and the *feeling–Object*.

Declaration

I herewith declare that I have produced this thesis without the prohibited assistance of third parties and without making use of aids other than those specified; notions taken over directly or indirectly from other sources have been identified as such. This work has not previously been presented in identical or similar form to any other Spanish or foreign examination board.

The thesis work was conducted from 2009 to 2016 under the supervision of Dr. D. Ricardo Sanz Bravo (Ph.D) and Dr. D. Ramón Galán López (Ph.D) at Universidad Politécnica de Madrid.

Madrid, 10th of February, 2016

M.G. Sánchez-Escribano, *Engineering Computational Emotion—A Reference Model for Emotion in Artificial Systems*, Cognitive Systems Monographs 33,
DOI 10.1007/978-3-319-59430-9

References

1. Modeling the Mechanisms of Emotion Effects on Cognition. In: Proceedings of the AAAI Fall Symposium on Biologically Inspired Cognitive Architectures, vol. 82–86 (2008)
2. Albus, J.S., Barbera, A.J.: Rcs: a cognitive architecture for intelligent multi-agent systems. Ann. Rev. Control **29**(1), 87–99 (2005)
3. Anderson, J.R., Bothell, D., Byrne, M.D., Douglass, S., Lebiere, C., Qin, Y.: An integrated theory of the mind. Psychol. Rev. **111**(4), 1036 (2004)
4. Andersson, J., Baresi, L., Bencomo, N., de Lemos, R., Gorla, A., Inverardi, P., Vogel, T.: Software engineering processes for self-adaptive systems. In: Software Engineering for Self-Adaptive Systems II, pp. 51–75. Springer (2013)
5. Arnold, M.B.: Emotion and personality (1960)
6. Ashby, W.R.: Principles of the self-organizing dynamic system. J. General Psychol. **37**(2), 125–128 (1947)
7. Ashby, W.R.: Principles of the self-organizing system. Principles of Self-organization, pp. 255–278 (1962)
8. Ashby, W.R.: Principles of the self-organizing system. In: Facets of Systems Science, pp. 521–536. Springer (1991)
9. Astrom, K.J., Murray, R.M.: Feedback Systems an Introduction for Scientists and Engineers. Hardcover (2008)
10. Atkinson, R.L., et al.: Hilgard's Introduction to Psychology, vol. 12. Harcourt Brace College Publishers, Philadelphia (1996)
11. Axelrod, R.M., Axelrod, R., Cohen, M.D.: Harnessing Complexity: Organizational Implications of a Scientific Frontier. Basic Books, New York City (2000)
12. Babaoglu, O., Jelasity, M., Montresor, A., Fetzer, C., Leonardi, S., van Moorsel, A., van Steen, M.: Self-star properties in complex information systems: Conceptual and practical foundations, volume 3460 of Lecture Notes in Computer Science (2005)
13. Bach, J.: The micropsi agent architecture. In: Proceedings of ICCM5, International Conference on Cognitive Modeling (2003)
14. Bard, P.: Emotion: I. the neuro-humoral basis of emotional reactions (1934)
15. Bartl-Storck, C., Dörner, D.: Comparing the Behaviour of psi with Human Behaviour in the Biolab Game (1998)
16. Bartneck, C.: Affective expressions of machine. Emotion **8**, 489–502 (2000)
17. Bates, J.: Virtual reality, art and entertainment. Presence **1**(1), 133–138 (1992)

M.G. Sánchez-Escribano, *Engineering Computational Emotion—A Reference Model for Emotion in Artificial Systems*, Cognitive Systems Monographs 33,
DOI 10.1007/978-3-319-59430-9

18. Bates, J., Loyall, A.B., Reilly, W.S.: An Architecture for Action, Emotion, and Social Behavior. Springer (1994)
19. Beck, A., Cañamero, L., Bard, K., et al.: Towards an affect space for robots to display emotional body language. In: 2010 IEEE Ro-man, pp. 464–469. IEEE (2010)
20. Bell, G. et al.: Lifelike computer characters: The persona project at microsoft. *Software Agent*, pages 191–222 (1997)
21. Bermejo-Alonso, J.: OASys: ontology for autonomous systems. Ph.D. thesis, Universidad Politécnica de Madrid (2010)
22. Bermejo-Alonso, J., Sanz, R., Rodríguez, M., Hernández, C.: An ontology–based approach for autonomous systems' description and engineering. In: Knowledge-Based and Intelligent Information and Engineering Systems, pp. 522–531. Springer (2010a)
23. Bermejo-Alonso, J., Sanz, R., Rodríguez, M., Hernández, C.: Ontology-based engineering of autonomous systems. In: 2010 Sixth International Conference on Autonomic and Autonomous Systems (ICAS), pp. 47–51. IEEE (2010b)
24. Blair, G., Bencomo, N., France, R.B.: Models@ run. time. Computer **42**(10), 22–27 (2009)
25. Bondi, A.B.: Characteristics of scalability and their impact on performance. In: Proceedings of the 2nd International Workshop on Software and Performance, pp. 195–203. ACM (2000)
26. Booch, G., Rumbaugh, J., Jacobson, I.: The Unified Modeling Language User Guide, 1999. Addison-Wesley Longman Inc, Boston (2010)
27. Boss, V.: Lecciones de matemática (2008)
28. Bower, G., Lazarus, R., LeDoux, J., Panksepp, J., Davidson, R., Ekman, P.: What is the relation between emotion and memory. In: Ekman, P., Davidson, R.J. (eds.) The Nature of Emotion: Fundamental Questions. Series in Affective Science, pp. 301–318 (1994)
29. Bower, G.H.: How might emotions affect learning. Handb. Emot. Mem.: Res. Theory **3**, 31 (1992)
30. Braitenberg, V.: Vehicles: Explorations in Synthetic Psychology (1984)
31. Breazeal, C.: Regulating human-robot interaction using emotions, drives, and facial expressions. In: Proceedings of Autonomous Agents **98**, 14–21 (1998)
32. Breazeal, C.: Emotion and sociable humanoid robots. Int. J. Human-Comput. Stud. **59**, 119–155 (2003)
33. Breazeal, C. et al.: A motivational system for regulating human-robot interaction. In: Aaai/iaai, pp. 54–61 (1998)
34. Brom, C., Pešková, K., Lukavskỳ, J.: What does your actor remember? towards characters with a full episodic memory. In: Virtual Storytelling. Using Virtual Reality Technologies for Storytelling, pp. 89–101. Springer (2007)
35. Brooks, R., et al.: A robust layered control system for a mobile robot. IEEE J. Robot. Autom. **2**(1), 14–23 (1986)
36. Brooks, R.A., Breazeal, C., Marjanović, M., Scassellati, B., Williamson, M.M.: The cog project: Building a humanoid robot. In: Computation for Metaphors, Analogy, and Agents, pp. 52–87. Springer (1999)
37. Brun, Y., Serugendo, G. D. M., Gacek, C., Giese, H., Kienle, H., Litoiu, M., Müller, H., Pezzè, M., Shaw, M.: Engineering self-adaptive systems through feedback loops. In: Software Engineering for Self-adaptive Systems, pp. 48–70. Springer (2009)
38. Bush, G., Luu, P., Posner, M.I.: Cognitive and emotional influences in anterior cingulate cortex. Trends Cognit. Sci. **4**(6), 215–222 (2000)
39. Cabibihan, J.-J., So, W.-C., Pramanik, S.: Human-recognizable robotic gestures. IEEE Trans. Auton. Mental Dev. **4**(4), 305–314 (2012)
40. Camras, L.A.: Expressive development and basic emotions. Cognit. Emot. **6**(3–4), 269–283 (1992)
41. Cañamero, D.: Modeling motivations and emotions as a basis for intelligent behavior. In: Proceedings of the first international conference on Autonomous agents, pp. 148–155. ACM (1997)
42. Cañamero, L.: Emotion understanding from the perspective of autonomous robots research. Neural Netw. **18**(4), 445–455 (2005)

43. Cannon, W.B.: The james-lange theory of emotions: a critical examination and an alternative theory. Am. J. Psychol. **39**, 106–124 (1927)
44. Carver, C.S., Scheier, M.F.: Origins and functions of positive and negative affect: a control-process view. Psychol. Rev. **97**(1), 19 (1990)
45. Cheng, B.H., De Lemos, R., Garlan, D., Giese, H., Litoiu, M., Magee, J., Muller, H.A., Taylor, R.: Seams 2009: Software engineering for adaptive and self-managing systems. In: Proceedings of the 2009 31st International Conference on Software Engineering: Companion Volume, pp. 463–464. IEEE Computer Society (2009)
46. Chong, H.-Q., Tan, A.-H., Ng, G.-W.: Integrated cognitive architectures: a survey. Artif. Intell. Rev. **28**(2), 103–130 (2007)
47. Church, A.: An unsolvable problem of elementary number theory. Am. J. Math. **58**, 345–363 (1936)
48. Cloutier, R., Muller, G., Verma, D., Nilchiani, R., Hole, E., Bone, M.: The concept of reference architectures. Syst. Eng. **13**(1), 14–27 (2010)
49. Conway, M.A.: Memory and the self. J. Mem. Lang. **53**(4), 594–628 (2005)
50. Conway, M.A., Pleydell-Pearce, C.W.: The construction of autobiographical memories in the self-memory system. Psychol. Rev. **107**(2), 261 (2000)
51. Cooper, L.: The rhetoric of aristotle (1933)
52. Cytowic, R.E.: Synesthesia: phenomenology and neuropsychology. Psyche **2**(10), 2–10 (1995)
53. Damasio, A.: Descartes' Error: Emotion, Reason, and the Human Brain. Grosset/Putnam, New York (1994)
54. Damasio, A.: The Feeling of What Happens: Body and Emotion in the Making of Consciousness. Harcourt (1999)
55. Damasio, H., Grabowski, T., Frank, R., Galaburda, A.M., Damasio, A.R., et al.: The return of phineas gage: clues about the brain from the skull of a famous patient. Science **264**(5162), 1102–1105 (1994)
56. Darwin, C.: The Origin of Species: By Means of Natural Selection or the Preservation of Favoured Races in the Struggle for Life. Murray, reissue edition (1859)
57. Darwin, C.: The Expression of the Emotions in Man and Animals, 3rd edn. Oxford University Press Inc, Oxford (2002)
58. Daubert, E., Fouquet, F., Barais, O., Nain, G., Sunye, G., Jézéquel, J.-M., Pazat, J.-L., Morin, B.: A models@ runtime framework for designing and managing service-based applications. In: 2012 Workshop on European Software Services and Systems Research-Results and Challenges (S-Cube), pp. 10–11. IEEE (2012)
59. Davidson, D.: Essays on Actions and Events: Philosophical Essays, vol. 1. Oxford University Press, Oxford (2001)
60. Davidson, R.J., Pizzagalli, D., Nitschke, J.B., Kalin, N.H.: Parsing the subcomponents of emotion and disorders of emotion: perspectives from affective neuroscience. In: Handbook of Affective Sciences, pp. 8–24 (2003a)
61. Davidson, R.J., Scherer, K.R., Goldsmith, H.: Handbook of Affective Sciences. Oxford University Press, Oxford (2003b)
62. Dey, A.K., Abowd, G.D., Salber, D.: A conceptual framework and a toolkit for supporting the rapid prototyping of context-aware applications. Human-Comput. Interact. **16**(2), 97–166 (2001)
63. Di Marzo Serugendo, G., Cheng, B., De Lemos, R., Giese, H., Inverardi, P., Magge, J., Andersson, J., Becker, B., Bencomo, N., Brun, Y., et al.: Software Engineering for Self-adaptive Systems: A Research Roadmap (2009)
64. Digman, J.M.: Personality structure: emergence of the five-factor model. Annu. Rev. Psychol. **41**(1), 417–440 (1990)
65. Downey, A.B.: Think Complexity: Complexity Science and Computational Modeling. O'Reilly Media, Inc, Sebastopol (2012)
66. Ekman, P.: Expressions and the nature of emotion. In: Scherer, K.S., Ekman, P. (eds.) Approaches to Emotion, pp. 319–343, Hillsdale, NJ: Erlbaum(1984)
67. Ekman, P.: Basic Emotions, pp. 45–60 (2005)

68. Ekman, Paul, Friesen, W.: Facial Action Coding System. Consulting Psychologist Press, Palo Alto (1978)
69. Elliot, C.: The affective reasoner: A process model of emotions in a multi-agent system. Northwestern University Institute for the Learning Sciences, Northwestern, IL (1992)
70. Elliott, C.D.: The Affective Reasoner: A Process Model of Emotions in a Multi-agent System (1992)
71. Ellsworth, P.C., Scherer, K.R.: Appraisal processes in emotion. Handb. Affect. Sci. **572**, V595 (2003)
72. Emde, R.N., Robinson, J., Corley, R.P., Nikkari, D., Zahn-Waxler, C., Emde, R., Hewitt, J.: Reactions to restraint and anger-related expressions during the second year, pp. 127–140. Genetic and environmental influences on developmental change, Infancy to early childhood (2001)
73. Ettinger, R.: Afterword from robert ettinger youniverse. The Philosophy of Robert Ettinger, p. 237 (2002)
74. Evans, E.: Domain-Driven Design: Tackling Complexity in the Heart of Software. Addison-Wesley Professional, Boston (2004)
75. Fernández-Abascal, E.G., Rodríguez, B.G., Sánchez, M. P.J., Díaz, M. D.M., Sánchez, F. J.D.: *Psicología de la emoción*. Editorial Universitaria Ramón Areces (2010)
76. Fouquet, F., Nain, G., Morin, B., Daubert, E., Barais, O., Plouzeau, N., and Jézéquel, J.-M.: An Eclipse Modelling Framework Alternative to Meet the Models@ Runtime Requirements. Springer (2012)
77. Fox, E.: Emotion Science: Cognitive and Neuroscientific Approaches to Understanding Human Emotions. Palgrave Macmillan, London (2008)
78. Franklin, S., Kelemen, A., McCauley, L.: Ida: a cognitive agent architecture. In: 1998 IEEE International Conference on Systems, Man, and Cybernetics, vol. 3, pp. 2646–2651. IEEE (1998)
79. Fridja, N.H., Kuipers, P., Ter Schure, E.: Relations among emotion, appraisal, and emotional action readiness. J. Pers. Soc. Psychol. **57**(2), 212–228 (1989)
80. Frijda, N.: The Emotions. Studies in emotion and social interaction. Cambridge University Press, Cambridge (1986)
81. Frijda, N.: The Laws of Emotion. Lawrence Erlbaum Associates (2007)
82. Frijda, N.H., Ortony, A., Sonnemans, J., Clore, G.L.: The Complexity of Intensity: Issues Concerning the Structure of Emotion Intensity (1992)
83. Frijda, N.H., Swagerman, J.: Can computers feel? theory and design of an emotional system. Cognit. Emot. **1**(3), 235–257 (1987)
84. Gardiner, H.-M., Metcalf, R.C., Beebe-Center, J.G. (1937). Feeling and Emotion: A History of Theories
85. Gardner, D.L., Lamberti, D.M., Prager, J.M.: Method and apparatus providing an intelligent help explanation paradigm paralleling computer user activity. US Patent 5,239,617 (1993)
86. Garlan, D.: Software architecture: a roadmap. In: Proceedings of the Conference on the Future of Software Engineering, pp. 91–101. ACM (2000)
87. Gebhard, P.: Alma: a layered model of affect. In: Proceedings of the Fourth International Joint Conference on Autonomous Agents and Multiagent Systems, pp. 29–36. ACM (2005)
88. Ghasem-Aghaee, N., Oren, T.: Towards fuzzy agents with dynamic personality for human behavior simulation. In: Summer Computer Simulation Conference. Society for Computer Simulation International, 1998, pp. 3–10 (2003)
89. Goertzel, B., Wang, P.: A foundational architecture for artificial general intelligence. Adv. Artif. Gen. Intell.: Concepts, Archit. Algorithms **6**, 36 (2007)
90. Goleman, D.: The emotional intelligence of leaders. Lead. Lead. **1998**(10), 20–26 (1998)
91. Goleman, D.: In: Kairós (ed.) Inteligencia Emocional (2012)
92. Graham, K., Knuth, D.E.: Patashnik, concrete mathematics. In: A Foundation for Computer Science (1989)
93. Gratch, J.: Emile: marshalling passions in training and education. In: Proceedings of the Fourth International Conference on Autonomous Agents, pp. 325–332. ACM (2000)

94. Gratch, J., Marsella, S.: A domain-independent framework for modeling emotion. Cognit. Syst. Res. **5**(4), 269–306 (2004)
95. Greenfield, J., Short, K., Cook, S., Kent, S.: Software Factories: Assembling Applications with Patterns, Models, Frameworks, and Tools. Wiley, New York (2004)
96. Gross, J.J., Thompson, R.A.: Emotion Regulation: Conceptual Foundations (2007)
97. Grossberg, S.: Competitive learning: from interactive activation to adaptive resonance. Cognit. Sci. **11**(1), 23–63 (1987)
98. Gruber, T.R.: A translation approach to portable ontology specifications. Knowl. Acquis. **5**(2), 199–220 (1993)
99. Haghnevis, M., Askin, R.G.: A modeling framework for engineered complex adaptive systems. IEEE Syst. J. **6**(3), 520–530 (2012)
100. Haken, H.: Advanced Synergetics. Springer, Berlin (1983)
101. Harper, D., MacCormack, D.: www.etymonline.com (2001)
102. Harré, R.: The Social Construction of Emotions. Blackwell (1986)
103. Hawes, N.: A survey of motivation frameworks for intelligent systems. Artif. Intell. **175**(5), 1020–1036 (2011)
104. Hawkins, J., Le Roux, S.: The Oxford Reference Dictionary. Oxford University Press, Oxford (1986)
105. Hayek, F.: The Sensory Order : An Inquiry into the Foundations of Theoretical Psychology. University Of Chicago Press, Chicago (1999)
106. Hernández Corbato, C.: Model-based Self-awareness Patterns for Autonomy. Ph.D. thesis, Industriales (2013)
107. Herrera, C., Sánchez-Escribano, M., Sanz, R.: The embodiment of synthetic emotion. Handbook of Research on Synthesizing Human Emotion in Intelligent Systems and Robotics, p. 204 (2014)
108. Higgins, E.T.: Self-discrepancy: a theory relating self and affect. Psychol. Rev. **94**(3), 319 (1987)
109. Hobson, J.A.: The Chemistry of Conscious States. Basic Books, New York City (1994)
110. Hornstein, H.A., LaKind, E., Frankel, G., Manne, S.: Effects of knowledge about remote social events on prosocial behavior, social conception, and mood. J. Persona. Soc. Psychol. **32**(6), 1038 (1975)
111. Hoyos, J.R., García-Molina, J., Botía, J.A.: A domain-specific language for context modeling in context-aware systems. J. Syst. Softw. **86**(11), 2890–2905 (2013)
112. Huhns, M.N., Jacobs, N., Ksiezyk, T., Shen, W.-M., Singh, M.P., Cannata, P.E.: Integrating enterprise information models in carnot. In: Proceedings of International Conference on Intelligent and Cooperative Information Systems, pp. 32–42. IEEE (1993)
113. Hurewicz, W., Wallman, H.: Dimension theory, vol. 107. Princeton, New Jersey (1941)
114. Isen, A.M., Simmonds, S.F.: The effect of feeling good on a helping task that is incompatible with good mood. Soc. Psychol. **41**, 346–349 (1978)
115. Izard, C.E.: Cross-cultural perspectives on emotion and emotion communication. Handb. Cross-Cultural Psychol. **3**, 185–221 (1980)
116. Izard, C.E.: Four systems for emotion activation: cognitive and noncognitive processes. Psychol. Rev. **100**(1), 68 (1993)
117. Izard, C.E.: Human Emotions. Springer Science & Business Media(2013)
118. James, W.: What is an emotion? Mind **9**(34), 188–205 (1884)
119. John, H.: Holland, Hidden Order: How Adaptation Builds Complexity (1996)
120. Johnson-Laird, P.N., Oatley, K.: Basic emotions, rationality, and folk theory. Cognit. Emot. **6**(3–4), 201–223 (1992)
121. Kauffman, S.A.: The Origins of Order: Self Organization and Selection in Evolution. Oxford university press, Oxford (1993)
122. Kent, S.: Model Driven Engineering. In: Integrated Formal Methods, pp. 286–298. Springer (2002)
123. Kephart, J., Chess, D.: The vision of autonomic computing. Computer **36**(1), 41–50 (2003a)

124. Kephart, J.O., Chess, D.M.: The vision of autonomic computing. Computer **36**(1), 41–50 (2003b)
125. Kim, S., Georgiou, P.G., Lee, S., Narayanan, S.: Real-time emotion detection system using speech: Multi-modal fusion of different timescale features. In: IEEE 9th Workshop on Multimedia Signal Processing, MMSP 2007, pp. 48–51. IEEE (2007)
126. Kleinginna, P., Kleinginna, A.: A categorized list of emotion definitions, with suggestions for a consensual definition. Motivat. Emoti. **5**(4), 345–379 (1981)
127. Klir, G.: An Approach to General Systems Theory. Van Nostrand Reinhold Co (1969)
128. Klüver, H., Bucy, P.C.: Psychic blindness and other symptoms following bilateral temporal lobectomy in rhesus monkeys. Am. J. Physiol. **119**, 352–353 (1937)
129. Koda, T., Maes, P.: Agents with faces: the effect of personification. In: 5th IEEE International Workshop on Robot and Human Communication, pp. 189–194. IEEE (1996)
130. Krevisky, J., Jordan, L.: Webster's Encyclopedic Unabridged Dictionary of the English Language (1996)
131. Laird, J.E., Newell, A., Rosenbloom, P.S.: Soar: an architecture for general intelligence. Artif. Intell. **33**(1), 1–64 (1987)
132. Lange, C.G.: The Emotions. Williams and Wilkins, Baltimore (1885)
133. Lange, C.G., James, W.: The Emotions, vol. 1. Williams & Wilkins(1922)
134. Langley, P., Choi, D., Shapiro, D.: A cognitive architecture for physical agents. Retrieved 28 Oct 2006 (2004)
135. Lazarus, R.S.: Psychological Stress and the Coping Process (1966)
136. Lazarus, R.S.: Thoughts on the relations between emotion and cognition. Am. Psychol. **37**(9), 1019 (1982)
137. Lazarus, R.S.: On the Primacy of Cognition (1984)
138. Lazarus, R.S.: Relational Meaning and Discrete Emotions (2001)
139. LeDoux, J.: Emotional networks and motor control: a fearful view. Prog. Brain Res. **107**, 437 (1996)
140. LeDoux, J.: Emotion Circuits in the Brain (2003)
141. LeDoux, J.E.: Emotional memory systems in the brain. Behav. Brain Res. **58**(1), 69–79 (1993)
142. LeDoux, J.E.: Emotion, memory and the brain. Sci. Am. **270**(6), 50–57 (1994)
143. LeDoux, J.E.: Emotion circuits in the brain. Ann. Rev. Neurosci. **23**, 155–184 (2000)
144. LEGO. Lego mindstorms
145. Leite, I., Martinho, C., Paiva, A.: Social robots for long-term interaction: a survey. Int. J. Soc. Robot. **5**(2), 291–308 (2013)
146. Leventhal, H.: A perceptual-motor theory of emotion. Adv. Exp. Soc. Psychol. **17**, 117–182 (1984)
147. Leventhal, H., Scherer, K.: The relationship of emotion to cognition: a functional approach to a semantic controversy. Cognit. Emot. **1**(1), 3–28 (1987a)
148. Leventhal, H., Scherer, K.R.: The relationship of emotion to cognition: a functional approach to a semantic controversy. Cognit. Emot. **1**, 3–28 (1987b)
149. Lewis, M.: The emergence of human emotions [w:]. In: Lewis, M., Haviland-jones, J., Barrett, l. (eds.), Handbook of Emotions, pp. 304–319 (2008)
150. Loia, V., Sessa, S.: Soft Computing Agents: New Trends for Designing Autonomous Systems, vol. 75. Springer (2001)
151. López Paniagua, I.: A foundation for perception in autonomous systems: Fundamentos de la percepción en sistemas autónomos. Ph.D. thesis, Escuela Técnica Superior de Ingenieros Industriales, Universidad Politécnica de Madrid (Spain) (2007)
152. Lowe, R., Ziemke, T.: The feeling of action tendencies: on the emotional regulation of goal-directed behavior. Front. Psychol. 2 (2011)
153. Loyall, B., Bryan, A., Bates, L.J.: Hap a Reactive, Adaptive Architecture for Agents (1991)
154. Macal, C.M., North, M.J.: Tutorial on agent-based modelling and simulation. J. Simul. **4**(3), 151–162 (2010)
155. MacLean, P.D.: The Triune Brain in Evolution: Role in Paleocerebral Fnctions. Springer Science & Business Media (1990)

156. Madl, T. et al.: Continuity and the Flow of Time: A Cognitive Science Perspective. Philosophy and Psychology of Time, 135–160. Springer International Publishing (2016)
157. Magee, C., de Weck, O.: Complex System Classification (2004)
158. Mandelbrot, B.B.: The Fractal Geometry of Nature, vol. 173. Macmillan (1983)
159. March, J.G., Simon, H.A.: Organizations revisited. Ind. Corp. Change **2**(1), 299–316 (1993)
160. Marsella, S.C., Gratch, J.: Ema: a process model of appraisal dynamics. Cognit. Syst. Res. **10**(1), 70–90 (2009)
161. McCrae, R.R., John, O.P.: An introduction to the five-factor model and its applications. Personal.: Critical Concepts Psychol. 60, 295 (1998)
162. McCulloch, W.: Embodiments of Mind (1965)
163. McGaugh, J.L.: Memory-a century of consolidation. Science **287**(5451), 248–251 (2000)
164. McLean, H.V.: The emotional health of negroes. J. Negro Educ. **18**, 283–290 (1949)
165. Mehrabian, A.: Framework for a comprehensive description and measurement of emotional states. Genetic Soc. Gen. Psychol. Monogr. **121**(3), 339–361 (1995)
166. Mehrabian, A.: Analysis of the big-five personality factors in terms of the pad temperament model. Aust. J. Psychol. **48**(2), 86–92 (1996a)
167. Mehrabian, A.: Pleasure-arousal-dominance: a general framework for describing and measuring individual differences in temperament. Curr. Psychol. **14**(4), 261–292 (1996b)
168. Militello, L.G., Dominguez, C.O., Lintern, G., Klein, G.: The role of cognitive systems engineering in the systems engineering design process. Syst. Eng. **13**(3), 261–273 (2010)
169. Miller, E.K., Cohen, J.D.: An integrative theory of prefrontal cortex function. Ann. Rev. Neurosci. **24**(1), 167–202 (2001)
170. Minsky, M.: Society of Mind. Simon and Schuster (1988)
171. Minsky, M.: The Emotion Machine. Pantheon, New York (2006)
172. Minsky, M.: The Emotion Machine: Commonsense Thinking, Artificial Intelligence, and the Future of the Human Mind. Simon and Schuster (2007)
173. Moffat, D., Frijda, N.H.: Where there's a will there's an agent. In: Intelligent Agents, pp. 245–260. Springer (1995)
174. Molina, J.J.G.: Desarrollo de software dirigido por modelos: conceptos, métodos y herramientas. Ra-Ma (2013)
175. Moors, A.: Flavors of appraisal theories of emotion. Emot. Rev. **6**(4), 303–307 (2014)
176. Moreno, R.A., Espino, A.L., De Miguel, A.S.: Modeling consciousness for autonomous robot exploration. In: Bio-inspired Modeling of Cognitive Tasks, pp. 51–60. Springer (2007)
177. Morris, J.S., Öhman, A., Dolan, R.J.: Conscious and unconscious emotional learning in the human amygdala. Nature **393**(6684), 467–470 (1998)
178. Müller, J.P.: Architectures and applications of intelligent agents: a survey. Knowl. Eng. Rev. **13**(04), 353–380 (1999)
179. Nareyek, A.: Review: intelligent agents for computer games. In: Computers and Games, pp. 414–422. Springer (2001)
180. Navon, D.: Forest before trees: the precedence of global features in visual perception. Cognit. Psychol. **9**(3), 353–383 (1977)
181. Neisser, U.: Cognitive Psychology, 1st edn. Prentice Hall, Upper Saddle River (1967)
182. Newell, A.: Physical symbol systems. Cognit. Sci. **4**(2), 135–183 (1980)
183. Newell, A.: Unified theories of cognition and the role of soar. In: Soar: A Cognitive Architecture in Perspective, pp. 25–79. Springer (1992)
184. Newell, A.: Unified Theories of Cognition (The William James Lectures). Harvard University Press, Massachusetts (2002)
185. Newell, A., Simon, H.A.: Computer science as empirical inquiry: symbols and search. Commun. ACM **19**(3), 113–126 (1976)
186. Nicolescu, G., Mosterman, P.J.: Model-Based Design for Embedded Systems. CRC Press, Boca Raton (2009)
187. Northoff, G.: From emotions to consciousness–a neuro-phenomenal and neuro-relational approach. Front. Psychol. 3 (2012)
188. Nwana, H.S.: Software agents: an overview. Knowl. Eng. Rev. **11**(03), 205–244 (1996)

189. Oatley, K.: Best Laid Schemes: The Psychology of the Emotions. Cambridge University Press, Cambridge (1992)
190. Oatley, K., Jenkins, J.: Understanding Emotions. Blackwells (1996)
191. Oatley, K., Johnson-Laird, P.: Cognitive approaches to emotions. Trends Cognit. Sci. **18**(3), 134–140 (2014)
192. Oatley, K., Johnson-Laird, P.N.: Towards a cognitive theory of emotions. Cognit. Emot. **1**(1), 29–50 (1987)
193. Oatley, K., Keltner, D., Jenkins, J.M.: Understanding Emotions. Blackwell publishing, (2006)
194. Ochsner, K.N., Bunge, S.A., Gross, J.J., Gabrieli, J.D.: Rethinking feelings: an fmri study of the cognitive regulation of emotion. J. Cognit. Neurosci. **14**(8), 1215–1229 (2002)
195. Ogata, T., Komiya, T., Sugano, S.: Development of arm system for human-robot emotional communication. In: 26th Annual Confjerence of the IEEE Industrial Electronics Society, 2000. IECON 2000, vol. 1, pp. 475–480. IEEE (2000)
196. Ogata, T., Sugano, S.: Emotional communication between humans and the autonomous robot wamoeba-2 (waseda amoeba) which has the emotion model. JSME Int. J. Ser. C Mech. Syst. Mach. Elem. Manuf. **43**(3), 568–574 (2000)
197. O'Reilly, R.C.: Biologically plausible error-driven learning using local activation differences: the generalized recirculation algorithm. Neural Comput. **8**(5), 895–938 (1996)
198. Ortony, A.: In: Trappl et al. (eds.) On Making Believable Emotional Agents Believable, pp. 189–211 (2002)
199. Ortony, A., Clore, G., Collins, A.: The Cognitive Structure of Emotions. Cambridge University Press, Cambridge (1988)
200. Panksepp, J.: The basics of basic emotion. In: The Nature of Emotion: Fundamental Questions, pp. 237–242. Oxford university press, New York (1994)
201. Papez, J.W.: A proposed mechanism of emotion. Arch. Neurol. Psych. **38**(4), 725–743 (1937)
202. Pérez, C.H.: Purposiveness and causal explanation
203. Pérez, C.H., Moffat, D.C., Ziemke, T.: Emotions as a bridge to the environment: on the role of body in organisms and robots. In: From Animals to Animats 9, pp. 3–16. Springer (2006)
204. Pessoa, L.: The Cognitive-Emotional Brain: From Interactions to Integration. MIT press, Cambridge (2013)
205. Picard, R.W.: Affective Computing (1995)
206. Picard, R.W.: What does it mean for a computer to "have" emotions. Emotions in Humans and Artifacts, pp. 87–102 (2003)
207. Picard, R.W., Picard, R.: Affective Computing, vol. 252. MIT press Cambridge (1997)
208. Plutchik, R.: A general psychoevolutionary theory of emotion. Theor. Emot. 1 (1980)
209. Poel, M., Akker, H., Nijholt, A., Kesteren, A.-J.: Learning emotions in virtual environments. Aust. Soc. Cybern, Stud (2002)
210. Posner, J., Russell, J.A., Peterson, B.S.: The circumplex model of affect: an integrative approach to affective neuroscience, cognitive development, and psychopathology. Develop. Psychopathol. **17**(03), 715–734 (2005)
211. Power, M., Dalgleish, T., et al.: Cognition and emotion: From order to disorder. Psychology press, Abingdon (2007)
212. Prinz, J.J.: Gut Reactions: A Perceptual Theory of Emotion. Oxford University Press, New York (2004)
213. Qingji, G., Kai, W., Haijuan, L.: A robot emotion generation mechanism based on pad emotion space. In: Intelligent Information Processing IV, pp. 138–147. Springer (2008)
214. Rank, S.: Affective Acting: An Appraisal-Based Architecture for Agents as Actors. na (2004)
215. Reeves, B., Nass, C.: Perceptual user interfaces: perceptual bandwidth. Commun. ACM **43**(3), 65–70 (2000)
216. Reilly, W.S.: Believable social and emotional agents. Technical report, DTIC Document (1996)
217. Reilly, W.S., Bates, J.: Building Emotional Agents (1992)
218. Robinson, C.: Dynamical Systems: Stability, Symbolic Dynamics, and Chaos. CRC Press, Boca Raton (1995)

219. Rohrer, B.: An implemented architecture for feature creation and general reinforcement learning. In: Workshop on Self-Programming in AGI Systems, Fourth International Conference on Artificial General Intelligence (2011)
220. Roseman, I.J.: Cognitive determinants of emotion: A structural theory. Rev. Personal.; Soc, Psychol (1984)
221. Roseman, I.J., Wiest, C., Swartz, T.S.: Phenomenology, behaviors, and goals differentiate discrete emotions. J. Personal. Soc. Psychol. **67**(2), 206 (1994)
222. Rosen, R.: On models and modeling. Appl. Math. Comput. **56**(2), 359–372 (1993)
223. Russell, S., Norvig, P.: Artificial Intelligence: A Modern Approach, 2nd edn. Prentice Hall, Upper Saddle River (2002)
224. Sage, A.P.: Conflict and risk management in complex system of systems issues. In: 2003 IEEE International Conference on Systems, Man and Cybernetics, vol. 4, pp. 3296–3301. IEEE (2003)
225. Salehie, M., Tahvildari, L.: Self-adaptive software: landscape and research challenges. ACM Trans. Auton. Adapt. Syst. (TAAS) **4**(2), 14 (2009)
226. Salovey, P., Mayer, J.: Emotional intelligence. Imagin. Cognit. Personal. **9**(3), 185–211 (1989)
227. Sánchez-Escribano, M., Sanz, R.: Emotions and the engineering of adaptiveness in complex systems. Proc. Comput. Sci. **28**, 473–480 (2014)
228. Sanz, R., Hernández, C., Sánchez-Escribano, M.G.: Consciousness, action selection, meaning and phenomenic anticipation. Int. J. Mach. Conscious. **04**(02), 383–399 (2012)
229. Sanz, R., López, I., Bermejo-Alonso, J., Chinchilla, R., Conde, R.: Self-x: The control within. In: Proceedings of IFAC World Congress (2005)
230. Sanz, R., López, I., Rodríguez, M., Hernández, C.: Principles for consciousness in integrated cognitive control. Neural Netw. **20**(9), 938–946 (2007)
231. Sanz, R., Sánchez-Escribano, G., Herrera, C.: A model of emotion as patterned metacontrol. Biol. Inspired Cognit. Archit. **4**, 79–97 (2013)
232. Schelling, T.C.: Micromotives and Macrobehavior. WW Norton & Company (2006)
233. Schemes, B.L.: The Psychology of Emotions. Cambridge University Press, Cambridge (1992)
234. Scherer, K.: Towards a dynamic theory of emotion: a component process approach. Approaches to Emotion, pp. 293–318. Laurence Erlbaum, Hillsdale, NJ (1987)
235. Scherer, K.R.: Emotion serves to decouple stimulus and response. The Nature of Emotion: Fundamental Questions, pp. 127–130. Oxford University Press, New york (1994)
236. Scherer, K.R.: Appraisal considered as a process of multilevel sequential checking. Apprais. Process. Emot.: Theory, methods, Res. **92**, 120 (2001)
237. Scherer, K.R., Schorr, A., Johnstone, T.: Appraisal Processes in Emotion: Theory, Methods. Research. Oxford University Press, New York (2001)
238. Selic, B.: The pragmatics of model-driven development. IEEE Softw. **20**(5), 19–25 (2003)
239. Simon, H.: Motivational and emotional controls of cognition. Psychol. Rev. **74**(1), 29–39 (1967)
240. Sloman, A.: Varieties of affect and the cogaff architecture schema. In: Proceedings of the AISB'01 Symposium on Emotions, Cognition, and Affective Computing. The Society for the Study of Artificial Intelligence and the Simulation of Behaviour (2001)
241. Sloman, A., Croucher, M.: Why Robots will Have Emotions (1981)
242. Smith, C.A., Kirby, L.D.: 4. consequences require antecedents. Feeling and thinking: The role of affect in social cognition, p. 83 (2001)
243. Smith, C.A., Lazarus, R.S.: Emotion and Adaptation (1990)
244. Smith, C.A., Lazarus, R.S.: Appraisal components, core relational themes, and the emotions. Cognit. Emot. **7**(3–4), 233–269 (1993)
245. Stahl, T., Voelter, M.: Model-Driven Software Development: Technology, Engineering, Management, 1st edn. Wiley, New York (2006)
246. Staller, A., Petta, P.: Towards a tractable appraisal-based architecture for situated cognizers. In: Grounding Emotions in Adaptive Systems, Workshop Notes, Fifth International Conference of the Society for Adaptive Behaviour (SAB98), Zurich, Switzerland (1999)
247. Sterling, P., Eyer, J.: Allostasis: A New Paradigm to Explain Arousal Pathology (1988)

248. Steunebrink, B. et al.: The Logical Structure of Emotions (2010)
249. Strongman, K.: The Psychology of Emotions: Theories of Emotions in Perspective. Willey, New York (1996)
250. Sun, R.: Duality of the Mind: A Bottom-Up Approach Toward Cognition. Psychology Press (2001)
251. Sun, R.: The clarion cognitive architecture: Extending cognitive modeling to social simulation, pp. 79–99. Cognition and Multi-agent, Interaction (2006)
252. Sun, R.: The motivational and metacognitive control in clarion, pp. 63–75. Modeling Integrated, Cognitive Systems (2007)
253. Swagerman, J.: The Artificial Concern Realization System Acres. A Computermodel of Emotion, Drukkerij Enschede (1987)
254. SysML, O.: Omg Systems Modeling Language (2006)
255. Sztipanovits, J., Karsai, G.: Model-integrated computing. Computer **30**(4), 110–111 (1997)
256. Taatgen, N.: The atomic components of thought. Trends Cognit. Sci. **3**(2), 82–82 (1999)
257. Taylor, R., Medvidovic, N., Dashofy, E.: Software Architecture: Foundations, Theory, and Practice, 1st edn. Wiley, New York (2009)
258. Thornton, R.K., Sokoloff, D.R.: Assessing student learning of newton's laws: the force and motion conceptual evaluation and the evaluation of active learning laboratory and lecture curricula. Am. J. Phys. **66**(4), 338–352 (1998)
259. Turing, A.: On computable numbers, with an application to the entscheidungsproblem. Proc. Lond. Math. Soc. s2-42(1):230–265 (1937)
260. Turing, A.: Computing machinery and intelligence. Mind **59**(236), 433–460 (1950)
261. Vashist, A., Loeb, S.: Attention focusing model for nexting based on learning and reasoning. In: BICA, pp. 170–174 (2010)
262. Velkquez, J.: Modeling Emotions and their Motivations in Synthetic Agents (1997)
263. Vogel, T., Giese, H.: Model-Driven Engineering of Adaptation Engines for Self-adaptive Software: Executable Runtime Megamodels, vol. 66. Universitätsverlag Potsdam (2013)
264. Vogel, T., Giese, H.: Model-driven engineering of self-adaptive software with eurema. ACM Trans. Auton. Adapt. Syst. (TAAS) **8**(4), 18 (2014)
265. Von Bertalanffy, L., et al.: The theory of open systems in physics and biology. Science **111**(2872), 23–29 (1950)
266. Wang, P.: Rigid Flexibility. Springer (2006)
267. Wang, Y.: On abstract systems and system algebra. In: ICCI 2006 5th IEEE International Conference on Cognitive Informatics, vol. 1, pp. 332–343. IEEE (2006)
268. Watson, D., Clark, L.A.: Emotions, Moods, Traits, and Temperaments: Conceptual Distinctions and Empirical Findings (1994)
269. Wierzbicka, A.: Why kill does not mean cause to die: the semantics of action sentences. Foundations of Language, pp. 491–528 (1975)
270. Wikipedia. Iteration — wikipedia the free encyclopedia
271. Wikipedia. Real number — wikipedia the free encyclopedia
272. Wikipedia. Upper ontology — wikipedia the free encyclopedia
273. Wirth, N.: Algorithms and Data Structures (1986)
274. Yu, F., Chang, E., Xu, Y.-Q., Shum, H.-Y.: Emotion detection from speech to enrich multimedia content. In: Advances in Multimedia Information Processing—PCM 2001, pp. 550–557. Springer (2001)
275. Zalta, E.N. (ed.) . The Stanford Encyclopedia of Philosophy. The Metaphysics Research Lab, Center for the Study of Language and Information, Stanford University Stanford, CA 94305-4115 (2012)
276. Zhang, D., Cao, C., Yong, X., Wang, H., Pan, Y.: A survey of computational emotion research. In: Intelligent Virtual Agents, pp. 490–490. Springer (2005)

Printed in the United States
By Bookmasters